MAN UNFOLDING

WORLD PERSPECTIVES

Volumes already published

I	Approaches to God	*Jacques Maritain*
II	Accent on Form	*Lancelot Law Whyte*
III	Scope of Total Architecture	*Walter Gropius*
IV	Recovery of Faith	*Sarvepalli Radhakrishnan*
V	World Indivisible	*Konrad Adenauer*
VI	Society and Knowledge	*V. Gordon Childe*
VII	The Transformations of Man	*Lewis Mumford*
VIII	Man and Materialism	*Fred Hoyle*
IX	The Art of Loving	*Erich Fromm*
X	Dynamics of Faith	*Paul Tillich*
XI	Matter, Mind and Man	*Edmund W. Sinnott*
XII	Mysticism: Christian and Buddhist	*Daisetz Teitaro Suzuki*
XIII	Man's Western Quest	*Denis de Rougemont*
XIV	American Humanism	*Howard Mumford Jones*
XV	The Meeting of Love and Knowledge	*Martin C. D'Arcy, S.J.*
XVI	Rich Lands and Poor	*Gunnar Myrdal*
XVII	Hinduism: Its Meaning for the Liberation of the Spirit	*Swami Nikhilananda*
XVIII	Can People Learn to Learn?	*Brock Chisholm*
XIX	Physics and Philosophy	*Werner Heisenberg*
XX	Art and Reality	*Joyce Cary*
XXI	Sigmund Freud's Mission	*Erich Fromm*
XXII	Mirage of Health	*René Dubos*

XXIII	Issues of Freedom	*Herbert J. Muller*
XXIV	Humanism	*Moses Hadas*
XXV	Life: Its Dimensions and Its Bounds	*Robert M. MacIver*
XXVI	Challenge of Psychical Research	*Gardner Murphy*
XXVII	Alfred North Whitehead: His Reflections on Man and Nature	*Ruth Nanda Anshen*
XXVIII	The Age of Nationalism	*Hans Kohn*
XXIX	Voices of Man	*Mario Pei*
XXX	New Paths in Biology	*Adolf Portmann*
XXXI	Myth and Reality	*Mircea Eliade*
XXXII	History as Art and as Science	*H. Stuart Hughes*
XXXIII	Realism in Our Time	*Georg Lukács*
XXXIV	The Meaning of the Twentieth Century	*Kenneth E. Boulding*
XXXV	On Economic Knowledge	*Adolph Lowe*
XXXVI	Caliban Reborn	*Wilfred Mellers*
XXXVII	Through the Vanishing Point	*Marshall McLuhan and Harley Parker*
XXXVIII	The Revolution of Hope	*Erich Fromm*
XXXIX	Emergency Exit	*Ignazio Silone*
XL	Marxism and the Existentialists	*Raymond Aron*
XLI	Physical Control of the Mind	*José M. R. Delgado, M.D.*
XLII	Physics and Beyond	*Werner Heisenberg*
XLIII	On Caring	*Milton Mayeroff*
XLIV	Deschooling Society	*Ivan Illich*
XLV	Revolution Through Peace	*Dom Hélder Câmara*
XLVI	Man Unfolding	*Jonas Salk*

WORLD PERSPECTIVES · *Volume Forty-six*

Planned and Edited by RUTH NANDA ANSHEN

MAN UNFOLDING

JONAS SALK

HARPER & ROW, PUBLISHERS
New York, Evanston, San Francisco, London

 For information address Harper & Row, Publishers, Inc., 10 East 53rd Street, New York, N.Y. 10022. Published simultaneously in Canada by Fitzhenry & Whiteside Limited, Toronto.

FIRST EDITION

STANDARD BOOK NUMBER: 06-013739-8

LIBRARY OF CONGRESS CATALOG CARD NUMBER: 74-181642

TO

THOSE WHO

GAVE ME

OPPORTUNITY

AND

ENCOURAGEMENT

Contents

World Perspectives

What This Series Means

It is the thesis of *World Perspectives* that man is in the process of developing a new consciousness which, in spite of his apparent spiritual and moral captivity, can eventually lift the human race above and beyond the fear, ignorance, and isolation which beset it today. It is to this nascent consciousness, to this concept of man born out of a universe perceived through a fresh vision of reality, that *World Perspectives* is dedicated.

My Introduction to this Series is not of course to be construed as a prefatory essay for each individual book. These few pages simply attempt to set forth the general aim and purpose of the Series as a whole. They try to point to the principle of permanence within change and to define the essential nature of man, as presented by those scholars who have been invited to participate in this intellectual and spiritual movement.

Man has entered a new era of evolutionary history, one in which rapid change is a dominant consequence. He is contending with a fundamental change, since he has intervened in the evolutionary process. He must now better appreciate this fact and then develop the wisdom to direct the process toward his fulfillment rather than toward his destruction. As he learns to apply his understanding of the physical world for practical purposes, he is, in reality, extending his innate capacity and augmenting his ability and his need to communicate as well as his ability to think and to create. And as a result, he is substituting a goal-directed evolutionary process in his struggle against environmental hardship for the slow, but effective, biological evolution which produced modern man through mutation and natural selection. By intelligent intervention in the evolutionary process man has greatly accelerated and greatly expanded the range of his possibilities. But he has not changed the basic fact that it remains a trial and error process, with the danger of taking paths that lead to sterility of mind and heart, moral apathy and intellectual inertia; and even producing social dinosaurs unfit to live in an evolving world.

Only those spiritual and intellectual leaders of our epoch who have a paternity in this extension of man's horizons are invited to participate in this Series: those who are aware of the truth that beyond the divisiveness among men there exists a primordial unitive power since we are all bound together by a common humanity more fundamental than any unity of dogma; those who recognize that the centrifugal force which has scattered and atomized mankind must be replaced by an integrating structure and process capable of bestowing meaning and purpose on existence; those who realize that science itself, when not inhibited by the limitations of its own methodology, when chastened and humbled, commits man to an indeterminate range of yet undreamed consequences that may flow from it.

Virtually all of our disciplines have relied on conceptions which are now incompatible with the Cartesian axiom, and with the static world view we once derived from it. For underlying the new ideas, including those of modern physics, is a unifying order, but it is not causality; it is purpose, and not the purpose of the universe and of man, but the purpose *in* the universe and *in* man. In other words, we seem to inhabit a world of dynamic process and structure. Therefore we need a calculus of potentiality rather than one of probability, a dialectic of polarity, one in which unity and diversity are redefined as simultaneous and necessary poles of the same essence.

Our situation is new. No civilization has previously had to face the challenge of scientific specialization, and our response must be new. Thus this Series is committed to ensure that the spiritual and moral needs of man as a human being and the scientific and intellectual resources at his command for *life* may be brought into a productive, meaningful and creative harmony.

In a certain sense we may say that man now has regained his former geocentric position in the universe. For a picture of the Earth has been made available from distant space, from the lunar desert, and the sheer isolation of the Earth has become plain. This is as new and as powerful an idea in history as any that has ever been born in man's consciousness. We are all becoming seriously concerned with our natural environment. And this concern is not only the result of the warnings given by biologists, ecologists and conservationists. Rather it is the result of a deepening awareness that something new has happened, that the planet Earth is a unique and precious place. Indeed, it may not be a mere coincidence that this awareness should have been born at the exact moment when man took his first step into outer space.

This Series endeavors to point to a reality of which scientific theory has revealed only one aspect. It is the commitment to this reality that lends universal intent to a scientist's most original and solitary thought. By acknowledging this frankly we shall restore science to the great family of human aspirations by which men hope to fulfill themselves in the world community as thinking and sentient beings. For our problem is to discover a principle of differentiation and yet relationship lucid enough to justify and to purify scientific, philosophic and all other knowledge, both discursive and intuitive, by accepting their interdependence. This is the crisis in consciousness made articulate through the crisis in science. This is the new awakening.

Each volume presents the thought and belief of its author and points to the way in which religion, philosophy, art, science, economics, politics and history may constitute that form of human activity which takes the fullest and most precise account of variousness, possibility, complexity and difficulty. Thus

World Perspectives endeavors to define that ecumenical power of the mind and heart which enables man through his mysterious greatness to re-create his life.

This Series is committed to a re-examination of all those sides of human endeavor which the specialist was taught to believe he could safely leave aside. It attempts to show the structural kinship between subject and object; the indwelling of the one in the other. It interprets present and past events impinging on human life in our growing World Age and envisages what man may yet attain when summoned by an unbending inner necessity to the quest of what is most exalted in him. Its purpose is to offer new vistas in terms of world and human development while refusing to betray the intimate correlation between universality and individuality, dynamics and form, freedom and destiny. Each author deals with the increasing realization that spirit and nature are not separate and apart; that intuition and reason must regain their importance as the means of perceiving and fusing inner being with outer reality.

World Perspectives endeavors to show that the conception of wholeness, unity, organism is a higher and more concrete conception than that of matter and energy. Thus an enlarged meaning of life, of biology, not as it is revealed in the test tube of the laboratory but as it is experienced within the organism of life itself, is attempted in this Series. For the principle of life consists in the tension which connects spirit with the realm of matter, symbiotically joined. The element of life is dominant in the very texture of nature, thus rendering life, biology, a trans-empirical science. The laws of life have their origin beyond their mere physical manifestations and compel us to consider their spiritual source. In fact, the widening of the conceptual framework has not only served to restore order within the respective branches of knowledge, but has also disclosed analogies in man's position regarding the analysis and synthesis of experience in apparently separated domains of knowledge, suggesting the possibility of an ever more embracing objective description of the meaning of life.

Knowledge, it is shown in these books, no longer consists in a manipulation of man and nature as opposite forces, nor in the reduction of data to mere statistical order, but is a means of liberating mankind from the destructive power of fear, pointing the way toward the goal of the rehabilitation of the human will and the rebirth of faith and confidence in the human person. The works published also endeavor to reveal that the cry for patterns, systems and authorities is growing less insistent as the desire grows stronger in both East and West for the recovery of a dignity, integrity and self-realization which are the inalienable rights of man who may now guide change by means of conscious purpose in the light of rational experience.

The volumes in this Series endeavor to demonstrate that only in a society in which awareness of the problems of science exists, can its discoveries start great waves of change in human culture, and in such a manner that these discoveries may deepen and not erode the sense of universal human community. The differences in the disciplines, their epistemological exclusiveness, the variety of historical experiences, the differences of traditions, of cultures, of languages, of the arts, should be protected and preserved. But the interrelationship and unity of the whole should at the same time be accepted.

The authors of *World Perspectives* are of course aware that the ultimate answers to the hopes and fears which pervade modern society rest on the moral fibre of man, and on the wisdom and responsibility of those who promote the course of its development. But moral decisions cannot dispense with an insight into the interplay of the objective elements which offer and limit the choices made. Therefore an understanding of what the issues are, though not a sufficient condition, is a necessary prerequisite for directing action toward constructive solutions.

Other vital questions explored relate to problems of international understanding as well as to problems dealing with prejudice and the resultant tensions and antagonisms. The growing perception and responsibility of our World Age point to the new reality that the individual person and the collective person supplement and integrate each other; that the thrall of totalitarianism of both left and right has been shaken in the universal desire to recapture the authority of truth and human totality. Mankind can finally place its trust not in a proletarian authoritarianism, not in a secularized humanism, both of which have betrayed the spiritual property right of history, but in a sacramental brotherhood and in the unity of knowledge. This new consciousness has created a widening of human horizons beyond every parochialism, and a revolution in human thought comparable to the basic assumption, among the ancient Greeks, of the sovereignty of reason; corresponding to the great effulgence of the moral conscience articulated by the Hebrew prophets; analogous to the fundamental assertions of Christianity; or to the beginning of the new scientific era, the era of the science of dynamics, the experimental foundations of which were laid by Galileo in the Renaissance.

An important effort of this Series is to re-examine the contradictory meanings and applications which are given today to such terms as democracy, freedom, justice, love, peace, brotherhood and God. The purpose of such inquiries is to clear the way for the foundation of a genuine *world* history not in terms of nation or race or culture but in terms of man in relation to God, to himself, his fellow man and the universe, that reach beyond immediate self-interest. For the meaning of the World Age consists in respecting man's hopes and dreams which lead to a deeper understanding of the basic values of all peoples.

World Perspectives is planned to gain insight into the meaning of man, who not only is determined by history but who also determines history. History is to be understood as concerned not only with the life of man on this planet but as including also such cosmic influences as interpenetrate our human world. This generation is discovering that history does not conform to the social optimism of modern civilization and that the organization of human communities and the establishment of freedom and peace are not only intellectual achievements but spiritual and moral achievements as well, demanding a cherishing of the wholeness of human personality, the "unmediated wholeness of feeling and thought," and constituting a never-ending challenge to man, emerging from the abyss of meaninglessness and suffering, to be renewed and replenished in the totality of his life.

Justice itself, which has been "in a state of pilgrimage and crucifixion" and now is being slowly liberated from the grip of social and political demonologies in the East as well as in the West, begins to question its own premises. The modern revolutionary movements which have challenged the sacred institutions of society by protecting social injustice in the name of social justice are here examined and re-evaluated.

In the light of this, we have no choice but to admit that the *un*freedom against which freedom is measured must be retained with it, namely, that the aspect of truth out of which the night view appears to emerge, the darkness of our time, is as little abandonable as is man's subjective advance. Thus the two sources of man's consciousness are inseparable, not as dead but as living and complementary, an aspect of that "principle of complementarity" through which Niels Bohr has sought to unite the quantum and the wave, both of which constitute the very fabric of life's radiant energy.

There is in mankind today a counterforce to the sterility and danger of a quantitative, anonymous mass culture; a new, if sometimes imperceptible, spiritual sense of convergence toward human and world unity on the basis of the sacredness of each human person and respect for the plurality of cultures. There is a growing awareness that equality may not be evaluated in mere numerical terms but is proportionate and analogical in its reality. For when equality is equated with interchangeability, individuality is negated and the human person transmuted into a faceless mask.

We stand at the brink of an age of a world in which human life presses forward to actualize new forms. The false separation of man and nature, of time and space, of freedom and security, is acknowledged, and we are faced with a new vision of man in his organic unity and of history offering a richness and diversity of quality and majesty of scope hitherto unprecedented. In relating the accumulated wisdom of man's spirit to the new reality of the World Age, in articulating its thought and belief, *World Perspectives* seeks to encourage a renaissance of hope in society and of pride in man's decision as to what his destiny will be.

World Perspectives is committed to the recognition that all great changes are preceded by a vigorous intellectual re-evaluation and reorganization. Our authors are aware that the sin of *hubris* may be avoided by showing that the creative process itself is not a free activity if by free we mean arbitrary, or unrelated to cosmic law. For the creative process in the human mind, the developmental process in organic nature and the basic laws of the inorganic realm may be but varied expressions of a universal formative process. Thus *World Perspectives* hopes to show that although the present apocalyptic period is one of exceptional tensions, there is also at work an exceptional movement toward a compensating unity which refuses to violate the ultimate moral power at work in the universe, that very power upon which all human effort must at last depend. In this way we may come to understand that there exists an inherent independence of spiritual and mental growth which, though conditioned by circumstances, is never determined by circumstances. In this way the great plethora of human knowledge may be correlated with an insight into the nature of human nature by being attuned to the wide and deep range of human thought and human experience.

Incoherence is the result of the present disintegrative processes in education. Thus the need for *World Perspectives* expresses itself in the recognition that natural and man-made ecological systems require as much study as isolated particles and elementary reactions. For there is a basic correlation of elements in nature as in man which cannot be separated, which compose each other and alter each other mutually. Thus we hope to widen appropriately our conceptual framework of reference. For our epistemological problem consists in our finding the proper balance between our lack of an all-embracing principle relevant to our way of evaluating life and in our power to express ourselves in a logically consistent manner.

Our Judeo-Christian and Greco-Roman heritage, our Hellenic tradition, has compelled us to think in exclusive categories. But our *experience* challenges us to recognize a totality richer and far more complex than the average observer could have suspected—a totality which compels him to think in ways which the logic of dichotomies denies. We are summoned to revise fundamentally our ordinary ways of conceiving experience, and thus, by expanding our vision and by accepting those forms of thought which also include nonexclusive categories, the mind is then able to grasp what it was incapable of grasping or accepting before.

In spite of the infinite obligation of men and in spite of their finite power, in spite of the intransigence of nationalisms, and in spite of the homelessness of moral passions rendered ineffectual by the technological outlook, beneath the apparent turmoil and upheaval of the present, and out of the transformations of this dynamic period with the unfolding of a world-consciousness, the purpose of *World Perspectives* is to help quicken the "unshaken heart of well-rounded truth" and interpret the significant elements of the World Age now taking shape out of the core of that undimmed continuity of the creative process which restores man to mankind while deepening and enhancing his communion with the universe.

RUTH NANDA ANSHEN

MAN UNFOLDING

I

Biology and Human Life

The purpose of this volume, by bringing together the knowledge we have about human life and about living systems generally, is to suggest a way of thinking about some of the burning issues of our time for which we seek solutions. This juxtaposition is intended to encourage us to look at human life from a biological viewpoint and to look, further and more deeply than has already been done, to biology for ideas relevant to human life.

I find it difficult to dismiss the idea that the basic question of our time, perhaps underlying all others, is a biological one. It concerns our understanding of the nature of man. Biological knowledge has developed to such a degree that it is now possible to interpret and to speculate about many aspects of human life from this point of view. Many of the facts of biology suggest models which might help us develop a more reasonable and realistic view of ourselves and our fellow man. Seeing ourselves in historical and dynamic perspective might serve as a basis for a common understanding that could reduce the destructive tendency of man toward man, and of man toward himself.

We start from the premise that life began on this planet when certain reactions occurred in inanimate material, involving the condensation of atoms of carbon, oxygen, hydrogen, and nitrogen to form molecules. In general, the properties of molecules are conferred by the architecture, or the relationship, of assembled atoms; some of the newly formed molecules possessed what we now call biological properties. This was when life began.

The earliest forms of life, from which human life eventually evolved, possessed the ability not only to survive but to make copies of themselves, and to change, or "improve," in ways that increased their chance of survival, and of replication, under a wide variety of conditions. At an early stage in the evolutionary process there had formed what has been referred to as "the thread of life"—the self-copying tape-like molecule that contains the code which transmits to each succeeding generation the information which, when decoded, forms the organism prescribed in it. This code contains the accumulated "wisdom" of previous generations of living things that survived having coped successfully with many changes in environment and many threatening influences.

The molecule containing the genetic code does not exist by itself. It is part of a cell which is an elaborate molecular machine possessing the unique property of self-copying, or self-replication, and the further property of differentiation into many different kinds of specialized cells, each of which performs a function of value to the organism that eventuates from the decoding process. The organism, and the cells and molecules that compose it, cannot be thought of separately from their environments as each evolved in a series of previous environments.

Environment is thought of here as the particular milieu in which the organism, or the cell, exists. Thus life in general, and human life in particular, must be thought of in the context in which it is found. The nature of the organism is defined as much by its surroundings as by the genetic code itself. The environment tests the organism's limit of adaptability and thus defines its character and reveals its potential for growth, development, and evolution.

Some time after the origin of molecules possessing properties compatible with life, a cell membrane was formed, in some way, thus creating an internal environment as distinct from the external environment. The molecules assembled within the cell membrane became organized, serving the purpose of survival and replication. The external factors essential for the survival of the organism so formed, and for its self-copying, are (a) the raw materials from which copies are made and (b) a source of energy for effecting the necessary reactions of synthesis.

Without going into the remarkable detail manifest in the "design" (pattern or order) of living things, it is clear that food and energy are required for the physicochemical machinery of life. The ultimate source of energy for living things on earth is the sun; organic materials, previously synthesized, contain solar energy in a combustible form which serves as fuel for organisms that feed upon such substances. Since the differences between man and other species are more often dwelt upon, recognition of their similarities will be helpful in understanding the problems that exist in the relationships between man and other living entities, and their environments, as well as the relationships among the individual organisms of a species, whether these be man or other living things. The economics of life with which man is confronted are not different in essence from those of any other living organism, no matter how simple, even a single cell. To cope with these problems, many biological "inventions" have developed over time, almost as if they had been designed for purposes that were anticipated. In reality, however, such biological "inventions"—as they might be thought of—were revealed only after the organism had been "challenged" by a test for survival under newly emerging conditions. Thus the path of evolution is the result of the interaction of the organism and its environment.

We know much in a general way about human life, and we know a great deal in particular about parts of the machinery of living organisms—including man. The lesson to be learned from the foregoing is that potentialities for dealing with the vicissitudes of life pre-exist but are never known until after the fact. Thus the capacity to survive cannot be known unless it is put to the test.

As we have said, traits are inherited from one generation to the next through the remarkable copying mechanism whereby the molecules which contain the hereditary information are replicated and passed on to succeeding generations. In addition, errors in the copying process occur, and further variety is introduced into living things through the process of mutation and selection.

Some mutations are evident at birth, while others do not become evident until exposure to an environment that evokes the expression of the newly arisen trait. The environment in this way evokes potential characteristics which may have either lifesaving value or an advantage in the competitive economics of life. However, the opposite may be just as true. Thus new properties may have positive or negative effects in relation to the demands of circumstances, and what may be advantageous for a time may later prove to be disadvantageous.

For the purpose of the present discussion, let us think of all manifestations of human life as having been latent in, and as having evolved from, prior forms. Let us also think of each as having persisted in the evolving genetic system and that environmental forces elicited the expression of the genetic potential.

In considering survival in the evolutionary process it is clear that at least two sets of judgments are involved. The first is related to the genetic changes that take place inside the germ cell, i.e., the egg or the sperm cell. When such changes occur, the response of the developing organism is that the somatic effects are either acceptable or unacceptable—in terms of organization of ordered function. If unacceptable, then the alteration does not reach the second point of judgment in intraspecies relations and in relation to the external environment. If acceptable to the organism and to the species, the further test of acceptability is by the environment, where conflicts occur in relation to other forms of life also in need of raw material and energy for survival and replication.

Thus, from the viewpoint of the internal organization of the organism itself, it is possible for a large reservoir of internal changes to accumulate in the "acceptable" category, to be available when needed at some later time, to cope with changes in environment and in ecologic relationships. It is also possible for there to be a loss, or a failure of development, of certain properties, which would result in an organism that is inadequate to cope with its environment or ecologically.

Therefore, there are two independent systems of selection, i.e., (a) internal, which pertain to the organism itself, and (b) external, which relate to effectiveness in coping with outside factors. In each of these the response to change may be acceptable or unacceptable. If unacceptable to either, i.e., internal or external factors, the effect is lethal; acceptability *by both* is required for survival. Adaptability reflects the ability of the organism and the "natural" biological and physical environment *to move in concert,* because both are in constant change and both are constantly interdependent, like the parts of an ecosystem, as if purposefully united. Thus purpose, or what seems to be purpose, if only for survival, influences the direction of growth, development, and evolution—whether in plants, animals, or man.

There is a counterpart to all of this in the way man functions. Ethically, man possesses a "conscience" that selects what is "right or wrong" for the individual himself, while the group makes a second judgment. The internal organization, or "conscience," of the individual must be satisfied, after which others judge the acceptability of the idea or of the action chosen. What is "right" and survives must have been acceptable, first, to the "nature" or "conscience" of the individual himself and then to "society," which consists essentially of many individuals who make up the group. This is a description of man's perpetual struggle to express his inner self, which must be exposed to judgment for acceptability by others as well as by the self.

Moral or ethical views established at any point in time need to be examined from a biological point of view, and a Biology of Man must take into consideration all aspects of his behavior and his life, i.e., as an individual and as a member of society and of the species.

It would seem from this that individual "change" is primary in evolution and that, through the process of "natural selection," "choices" are then made for survival. If we think of social or cultural evolution in this way, then "innovation" by individuals followed by "choice" serves as the equivalent, in the "social selection" process, of mutation and selection in the process of "natural selection." Since the "individual" exercises judgment in relation to "social" (i.e., the equivalent of "natural") selection, he can, to this extent, influence "cultural" evolution by his choices. The choices of individuals then influence the choices of others as well. The "values" on the basis of which "judgments" and "choices" are made from among the options that exist then determine the direction of evolution, whether biological (for survival) or cultural (for satisfaction in life).

The individual and the group operate together in determining the future. The individual plays a decisive primary role, and the group operates either by limiting or by not inhibiting the influence of the individual. This is the way of biological evolution and seems also to be true for the social and cultural evolution of man.

2

The Biological Way of Thought

Man is actively involved in change and evolution. Through his choices, through the way in which he thinks, through what he thinks about, and through what he does, he has the capability of facilitating either evolution or devolution.

As more and more is revealed about the fine structure of the organized complexity that composes man, biological knowledge and thinking will be of increasing usefulness to scientists who think about man as well as to philosophers or others concerned with evolving and unfolding man.

Cause and effect exist in living systems in general and in man, as well as in the physical universe, but not everything is predictable. And we would like to know what can be predicted and to what extent. This kind of question, when applied to realms of thought heretofore the domain of philosophers, stimulates scientific inquiry. We are proposing the use of a "theoretical-experimental way of thought" as distinguished from a "philosophical-speculative way of thought" for dealing with questions in the human realm. The following are illustrations of the "thoretical-experimental way of thought": The phenomenon of evolution had been recognized by many, other than Darwin, from observations generally available. But it was Darwin who suggested how evolution works and who conceived of the idea of natural selection and brought evidence to bear in support of his hypothesis. Similarly, Watson and Crick conceived of the self-copying mechanism by which hereditary information is passed on to succeeding generations.

Material that comes from philosophy, history, and the arts, as well as from physics, chemistry, and biology, can be treated according to the "theoretical-experimental" way of thought. New concepts will arise from both sources and new ideas will emerge from relating each to the other.

Ideas are themselves substantive entities with the power to influence and even to transform human life. In effect, ideas are not unlike food, vitamins, or vaccines. They evoke inherent potential for growth and development and can affect the course of evolution. They often produce unpredictable effects, leading to new experiences which lead to still further unforeseeable effects.

While we need concepts that are appropriate and helpful in understanding and doing something about the human condition, and about individual human lives, biologists know that consequences far in the future can no more be imagined than the emergence of man could have been predicted early in evolutionary time. Although a great deal is already known about what might be called biological mechanisms and processes, much more remains to be learned about psychosocial processes and mechanisms that arise from the bioanthropological unit we call man. Many of man's attributes still need to be explained, such as his aesthetic sense and its expressions and those transcendental qualities which might be referred to as the art in him—the essence of his character and personality that distinguishes each individual from all others. Is this a proper subject for scientific inquiry? How might this be done and by whom? At the present time there are many who are interested in knowledge of this kind. They are asking: What is man for? What does he produce or create? Where does he fit into the scheme of living things? Where does he fit in his own species? What are his strivings? What does he seem to want? What is it that seems to give him contentment and satisfaction?

Desire is an instinctual force which propels man to experience, discover, and test extremes. Reason, based on knowledge and memory, acts concurrently as a moderator. It is necessary to understand the action and reaction of these two forces, whether they are innate or acquired, and in what way they develop. Man often becomes a battlefield when these opposing forces confront each other. To what extent reason and desire can be harmoniously related has been the problem with which man has been coping with incomplete success since his beginning. Can knowledge of the nature and structure of living systems, from which man has evolved, help him meet the challenge and achieve balance between desire and reason?

The problem we face in many aspects of life is in not knowing what we want. In other aspects the difficulty is in knowing and not being able to attain. Since desire is often more compelling than reason, it is important and necessary for man to know and manage his desire. To what extent, then, can he through reason satisfy his desire, or divert, thwart, or postpone its fulfillment—if that is his wish?

Just as we now comprehend the embryologic development of cells and organs, so we should like to study the development of those human qualities that distinguish human life from other forms of life. We could then evaluate not only the factors responsible for gross differences in the intellectual and intuitive realms but also the factors related to the fine and subtle differences in the stages and moods in each of our lives.

It would be of further interest to understand the origin and nature of motivation and attitude, of what goes into their construction. It is not enough to say that these are subjective and therefore beyond our capacity to study; motivation and attitude can be measured objectively by behavior. To what extent does conscious will play a role, and to what extent is this genetically determined? What effects are established at the moment of conception and what subsequent influences also contribute to their formation?

The extent to which man has so successfully probed the nature of living matter and has unraveled the structure of nucleic acids, proteins, and other complex molecules makes it seem likely that his curiosity and ingenuity will someday contribute, in equal measure, to understanding the special qualities of his own being as well as the molecular make-up of his own body.

From the foregoing it is apparent that as a physiological, psychological, and social entity—in an environment which he has created, and which he is polluting physically, psychologically, and socially—man has come to the fore as a unit, the unit being man *and* his environment—with his brain the essential organ that must be understood in all its manifestations. Thus man the individual *and* mankind the organism in an environment must be looked upon as a unit, and must be examined with the eyes and minds of scientists working together with thinkers from other disciplines.

To understand and deal with mankind as an organism, new kinds of scientists are needed and new kinds of thinkers—those trained not only in the conventional disciplines but trained also to address themselves to questions related to the many interrelated facets of human existence. A new form of scientific and cultural education will be required as it becomes clearer that man is partly a physicochemical machine and partly a *being* alive in the cosmos.

In historical perspective, man's place on earth in relation to the origin of the universe is as a second of time in a play depicting events from then until now. Depending upon the choices he makes, his future on earth will be either longer or shorter than his past. When we make this statement, our orientation to the future—which is a well-developed trait in man—is stimulated to imagine what life may be like in time to come.

The switch from the view that man is unchanging to the view that human life evolves, not only biologically but socially and culturally, is now generally accepted. We now constantly imagine what it might be like if things were different. Such dynamic thoughts lead to change. Part of the great unrest and upheaval in the world today is due not so much to change itself as to the increased speed of change. The opportunity for change also affects the liberation of nations, and the freedom of individuals, from religious, cultural, social, and economic constraints which have been regarded by many as unchanging aspects of human life.

At one time or another in our own existences we are tempted to take an active part in changing life in order to create a better world in which to dwell. Each man attempts to do so in his own way and from his own point of view. But what is now important is a common understanding of the questions, the problems, and the issues, even if a common point of view as to the avenues to solution has not yet been achieved.

Although man's physical evolution has its own natural pace, man has so accelerated his cultural evolution as to make it seem that physical and cultural evolutionary processes are now taking place at intolerably different rates.

Man evolved physically as a result of genetic-somatic changes, with the survival of those which best fitted the prevailing circumstances. Man's fitness for survival has been amply tested under circumstances heretofore dictated by nature. Now, however, man creates the circumstances in which he finds himself. Thus a new kind of awareness arises from the increased pressure and stress of change that he has himself created. By increasing the variety of opportunities for choice, he has accelerated his cultural evolution to a point at which the dominant influence is change, or the speed of change, itself.

All this can be verified by our own senses. We can all testify that each of us has had to develop ways of coping with the tempo of life and the speed of change, just as man in the past had to cope with other kinds of change that threatened his existence or his integrity.

In a world in which change has accelerated from its natural tempo to one in which man has made change the order of the day, change itself has become man's principal problem. A "new man" is now emerging. The natural selective pressure will now favor one who not only accepts change but welcomes it and contributes to it. But having made this statement, we must acknowledge the need for judgment in determining the direction change will take and in making choices in the use of limited resources. Since all will not agree at all times, the inevitability of conflict and the need for coping with it become immediately apparent.

In the past, by and large, man had to strive for things he did not possess. While this is still true in large segments of the world, in others the problems posed are those of excess. Whether man has to cope with excess or with insufficiency, his physical and mental health are in jeopardy—and it is to the health of his body and of his mind that his attention is now drawn.

Man must ask himself whether or not, and then how, he can possibly bring about more reasonable control in respect to either excess or insufficiency. When passivity allows deprivation to persist, or when activity leads to excess, what, then, is man's role in relation to himself? To other men? To nature?

Just as each past age has had its particular difficulties and diseases, and their associated causes, the present age has developed a syndrome as real and recognizable as those caused in times past by the tubercle bacillus, the spirochete of syphilis, the flea and the bacillus of plague, the mosquito and the plasmodium of malaria, the diphtheria bacillus, or deficiencies of vitamin C in scurvy and of vitamin D in rickets, for example. The distinction that can be made between man's new syndrome and the nature of many of his past maladies is that previously his problems arose from causes attributable primarily to factors outside himself. Now many of his major problems seem to arise from within himself. Man, accustomed to reducing the incidence of disease by improvements in sanitation, in water and food supplies, by the development of preventives and treatments administered as pills or injections, seems to expect the same kind of answers for all his difficulties, when, in fact, the cure for some of his present imbalance can come only from something educed from within himself. Man must look inside, as well as outside, himself for the remedy of the problems with which he is confronted. We find that at times he does not know what to do; at other times he knows but does not do it; and at still other times he does the right thing without knowing the reason.

For a balanced approach to comprehending the human condition, there is need for the concerted application of the methods of thought of the biologist, who works at the molecular, cellular, and organismic level; the philosopher, who is concerned with ontological values; and the physician, who deals with human health in all its aspects. Each must relate deeply to his own special interests, and together they need to be concerned with their common interests.

How can this be accomplished? All work—and especially new work—is done by those who are inspired, by those who have what Leo Szilard once described as "the divine spark." This inspiration must be recognized and encouraged. It is different from the burning embers of a dying fire. It is a spark that will ignite a great flame of understanding and release great power to change life in ways toward which man's hopes have long been directed.

But let there be no misunderstanding. What is required are men who can think alone *and* with others, at the same time receiving encouragement from those who can implement their newly generated thoughts. Above all, emphasis must be placed upon motivation and attitude as well as upon intellectual ability. It is as if the very product of their inquiry must be available, for use upon themselves, before they can begin to extend their probings deeply and broadly enough for them to be of general value.

The distance between this broad statement and a concrete step may be long in coming and difficult of accomplishment unless we are alert to the discovery of the natural talent that can conceive the way such problems can be approached. A combination of imagination and the capacity to act with skill and technical competence is necessary. It is from the generation on the brink of making commitments for the first time, or from those who have the opportunity to make new commitments, in a second or third career, that the talent and wisdom will come, to provide insight into those of man's highly developed attributes that define his humanness and that can be studied to only a limited degree in other forms of life.

At present those whose pursuits are intellectual, creative, or skilled, and for whom the problems of food, shelter, and mobility have been solved by technological means, will find it necessary to address themselves to a new set of questions. The adversity that served to orient man in nature has been largely replaced by the adversity inflicted by man upon himself.

Each age has been characterized by human achievements which have profoundly affected the interrelationships of beings on this planet. The result has often been expressed in an increased number and distribution of people over the surface of the earth. Not only the number, but also the quality, of human lives has changed.

A crisis seems to be developing in the "advanced" segments of the world—perhaps most prominently in the more advanced segments of the so-called Western world, but not, by any means, limited to it. The circumstance with which the crisis seems to be associated is one of plenty, in which the major struggles for existence and for survival have been met and the purpose of life has become a question consciously posed. We can now choose ways of life from a number of alternatives.

The problem of choice strikes especially during adolescence, when individuals begin to assume responsibility for themselves, and again at other times when the circumstances of life change, whether due to external or internal causes, and new choices must be made. It becomes apparent that the problem originates, in part at least, in early life, as the personality is being constructed under the influence of its earliest experiences during the most impressionable part of the life cycle.

The distinguishing character of the present age of man and his diverse subcultures has recent as well as remote causes. To what extent will an awareness of these causes help man individually or collectively in dealing with his immediate and future problems?

As we pose questions such as these, we become conscious of many things. We become conscious of the enormous advances that have occurred in knowledge and technology. We become conscious of the role played in human evolution not only by man's erect posture and his opposable thumb but by his improbable and remarkable brain. Who could have foreseen the evolution of the human brain from the forms of life that emerged from the primordial ooze in which life began? And yet it indeed happened.

It is clear that man's brain demands fuller understanding. Of all the organs in his body, it may be said to be the one which contains his self, and, in a way, it is the organ which all other organs serve. Man's relationship to his self and to others—his behavior as an individual or as a social being—is all determined, ultimately, by the way in which his brain functions. We are not far away from the moment when this will be recognized as our major preoccupation. For some, that time has already come. Those who are now devoting themselves to such studies are in the vanguard of thinkers and workers in their respective scientific and scholarly fields—whether they are biologists or others among whom we include psychologists or psychobiologists; sociologists or social biologists; physicians, philosophers, or even cosmologists concerned with the nature, organization, and complexity of matter including the complexity of man.

3

Analogies Between Immunologic and Psychologic Phenomena

In the course of his evolution man coped with deprivation, disease, and insufficiency by trying to determine the causes and cures of his wants—of things he lacked yet desired. He did this by replacing belief with knowledge through a pursuit called science.

Thus science was born when, in response to needs and wants, substance was given to flashes of intuition by those who may be said to have been the first to practice the art of scientific inquiry and proof. Science is, in a way, a human activity which was first practiced as an art. Its power was soon recognized and began to be used not only to give reality to intuitive ideas but as a way of consciously posing questions.

Until the dawn of modern times, man was subject to plagues and pestilences for reasons that he could not understand. For example, at the end of the eighteenth century a discerning individual recognized, by chance, the connection between two seemingly unconnected events—the clear complexion of the milkmaid and her exposure to cowpox. This led Jenner to discover a way to prevent smallpox. From that beginning there developed, by rapid acceleration, as we now measure the scale of human history, the evolution in the knowledge and the prevention of infectious diseases. Thus Jenner initiated a revolution, carried on by others who made connections between seemingly unrelated facts and thereby brought about a qualitative change in the life of man.

Another discovery revealed by relating the seemingly unrelated occurred in the field of nutrition. Scurvy was the scourge of navigators and navies until it was observed that it did not occur among seamen who ate limes. This led not only to the prevention of scurvy, by the addition of citrus fruits to the diet, but contributed to the discovery of vitamins, which, together with vaccines, have become part of our present natural way of life.

Thus in the course of time man has learned how to prevent or treat many of the diseases and pestilences that prevailed in the past, and has changed his life by bringing under control the factors arising outside his own body which caused them. For the most part, the diseases, or the disorders of excess or insufficiency, which still plague him arise from internal rather than external causes.

As man has turned his attention inward in an effort to understand the nature of the influences within himself, he has come face to face with an order of complexity far greater than any he has tried to encompass heretofore.

It would not help to emphasize this complexity here. We would prefer to try to simplify the problem by indicating the nature of the basic relationships that exist in biological systems, in order to understand better how a cell works and, from this, to grasp how the whole organism functions. We will try to reveal the meaning which such understanding may have for human life, now that we have become increasingly conscious of our need to deal with problems arising from within ourselves.

The body is made up of cells. Each organ has a characteristic cell. The liver cell is different from a muscle cell or a brain cell. Moreover, there is more than one kind of cell in each organ, each serving different purposes, so that, for example, the liver or kidney is made up of many different kinds of cells. Then there are cells that wander and serve scavenger functions. This kind of specialization has taken place in the course of evolution through the persistence of systems indispensable for survival.

At this point let us look at the meaning of the revolutionary biological discovery that the cell is made up of functionally distinct molecules. The compactness of the cell and the efficiency with which it functions, and the significance of subtle molecular differences, is remarkable to behold.

For almost two decades it has been known that hereditary information is coded in a molecule made up of a variable sequence of four elements. Related molecules from two parents interact, pair, separate, and make copies which are then represented in each new cell as cell division proceeds from a single fertilized egg. As the environment of each cell changes by virtue of the increase in the number of cells in the enlarging mass, ordered commitments take place resulting in cell differentiation and specialization. Thus the single fertilized cell possesses at the beginning the capacity to be any kind of a cell, and although each cell has its share of this potential, there comes a point in cell division in the developing embryo at which commitment occurs and the nature of the cell becomes defined for its lifetime. When the flow of genetic information departs from the "prescribed" order, malformation, disease, or death may ensue. In time, all cells age, and at a critical point the death of the organism occurs. This is a way of looking at birth, growth, development, and death.

The complex internal machinery required to carry on the particular function of each specialized cell must be under precise control; raw materials need to be converted into essential elements for use within the cell or elsewhere in the body by other cells. Apart from the regulatory systems in the cell and in the organism engaged in maintaining order and constancy in the internal environment, there are whole systems of cells concerned primarily with adaptation to the external environment. Some environmental influences are harmful and must be recognized and dealt with appropriately. The cells that deal with the external environment in this way are those of the nervous system and of the immunologic system.

For example, there is the immunologic system which evolved to protect the organism from foreign invaders and from foreign bodies of various kinds. This system has evolved in many ways to maintain the integrity of the organism. Many different kinds of antibody, from the structural and functional point of view, have developed to recognize and protect the organism against harmful invaders. At times, the immunologic system itself has harmful effects, as, for example, when it fails to distinguish self from not-self and produces an antibody that is destructive to the body's own tissues, as in the so-called autoimmune or auto-allergic diseases. In the same way, it may be that an antibody-antigen complex that seemingly evolved to protect the fetus during pregnancy—since the fetus is, in a way, a foreign body in the mother—sometimes contributes adversely by protecting cancerous tissues which appear to evoke a kind of "foreign body reaction" similar to the embryo and the fetus.

The immunologic system may be said to react "instinctively" to influences experienced as harmful in past evolutionary time. But it is not perfect in its discrimination and possesses the potentially inherent danger of turning against self tissue.

Thus the immunologic system has evolved with both positive and negative values. Its controlled suppression or activation is one of the most challenging problems in present-day experimental biology.

We see parallels to the immunologic system in the central nervous system. The nervous system also reacts "instinctively" to stimuli that arise internally as well as externally. Another parallel between the immunologic system and the nervous system is that at birth both are partly developed ("instinctive" behavior) and at the same time are capable of further development ("learned" behavior). It is interesting to note the similarity in language used to identify phenomena associated with each system. For example, the "conditioned reflex" of psychology, which implies a process of learning and of memory, may correspond functionally to what is referred to as "immunologic conditioning," which is at the basis of the reaction of recall, or the booster response, and, in a way, resembles the conditioned reflex.

There is also the phenomenon known as immunologic tolerance and another known as immunologic rejection. We need not explain these in further detail at this moment, other than to let the words themselves convey the idea.

The phenomenon of immunologic allergy is well known. We also are aware that people sometimes react to each other as if they were allergic—a term that is used probably as much in common speech as in professional terminology.

We speak of the uniqueness of the individual in psychological terminology, and there is the counterpart to this in immunology. There are a number of other terms that provide additional justification for considering these two systems side by side.

Additional justification for a comparative consideration of the organism's immunologic and psychologic functions is provided by thoughts about their survival value for man. The presence of both functions implies the necessity for their existence. This is obviously true for all the functions present in any living organism. Both systems serve essentially similar purposes, each in different ways, playing particularly prominent roles in adaptation to the external environment, although they also play important roles in maintaining the equilibrium of the internal environment of the organism. Thus, both the immunologic system and the central nervous system, prominent in protecting the organism against externally originating harmful influences, are *special* organs for adaptation, if this can be said of any organ.

In the simplest organism, consisting of but a single cell, all the vital functions are performed by the one cell. However, as greater complexity developed, in organisms consisting of more than one cell, divisions of responsibility evolved, along with greatly increased specialization. This can be illustrated, for example, by the division of responsibility for nutrition between the respiratory and gastrointestinal organs. The respiratory tract is concerned with the intake of gaseous nutrients and the gastrointestinal tract with liquid and solid nutrients. Similarly, the hepatic and renal functions both eliminate waste products or convert toxic substances to harmless ones and, in this way, help maintain homeostasis. These are oversimplified examples to illustrate a much more complex process. The point we are making is that the immunologic system and the central nervous system seem to be two different aspects of a system-complex which protects the organism and facilitates adaptation.

If the immunologic and central nervous systems are thought of in parallel, and if the analogies proposed are not too far-fetched, it is suggested that, in a limited way at least, certain phenomena of the immunologic mechanism could serve as a model for understanding certain functions of the nervous system.

To develop this line of reasoning, it would be necessary to imagine whether or not the immunologic system has counterparts equivalent to both instinctive behavior and learned behavior. For instance, can we regard as an example of "immunologic instinct" the phenomenon of "natural immunity" through which the organism from the moment of birth, without any postnatal experience, exhibits the capacity for dealing with microorganisms such as those that exist, without evident harm, on the skin, in the pharynx, or in the intestinal tract? Under some circumstances these same organisms can and do exert pathogenic effects, as in other species and in the poorly controlled diabetic. This is what occurs antemortem, when the "natural defenses" break down generally and invasion by otherwise nonpathogenic organisms ensues.

Thus "natural immunity" and "instinct" might be considered analogous—and the "acquired immunity" of the body, which is a learned response, might be considered as the analogue of the "learning capacity" of the mind. At the moment we are deferring consideration of any property of the immunologic system that corresponds to the anticipatory, the creative, and the integrative functions of the brain.

Let us examine a few of the immunologic phenomena that have prompted these speculations.

The psychological uniqueness of all humans is self-evident. At the same time it is obvious that all people look different and are different in other ways. The immunologic uniqueness of individuals is readily demonstrable from attempts at skin and organ transplantation. It is not possible successfully to graft skin from one individual to another, or to transplant an organ from one person to another unless the individuals involved are identical twins or artificial immunosuppression has been applied. While identical twins may be similar immunologically, we also know that there are differences between them in behavior and personality. Thus there seems to be a difference in the order of magnitude of likeness. The *person* is unique even though he may be remarkably similar, in certain major respects, to another individual derived from a single fertilized ovum.

The development of differences is simple to understand. The multiplicity of genes and the combinations that are possible from the union of two sets of chromosomes, which occurs in sexual reproduction, are so great that identity is approximated only when the same fertilized ovum divides to give rise to two separate individuals. However, even in such instances, postgenetic influences, including psychologic, can alter the relationships of individuals who are identical twins.

Thus individual differences occur through the operation of genetic laws. But, by certain manipulations, these laws, which govern phenomena which tend to increase such differences, can be utilized to reduce them. For example, for studies requiring animals that are essentially identical immunologically, inbred strains have been developed by brother-sister matings, which approximate such effects after a sufficient number of generations of such matings. In many strains of mice and rats and other laboratory animals, such carefully controlled inbreeding has produced sufficient diminution of differences as to result in individuals which are essentially identical immunologically. Among such animals skin grafts and organs can be exchanged and tumors can be transplanted. Thus each individual can be regarded as interchangeable with another so far as the genetically determined immunologic composition of the tissues is concerned. There are also genetically determined individual differences in reactivity, or in responsiveness, of the different components of the immunological system which can be exaggerated by inbreeding.

An animal does not normally develop antibodies to its own tissues because at some point in embryonic or fetal development a mechanism for distinguishing self from not-self comes into play. Whatever proteins, or other antigens, are present at that moment are generally regarded henceforth as representing "self," and any others that may enter at some later time are then regarded as "not-self." Thus the introduction of a foreign protein or of foreign tissue, in the form of a microorganism, whether it be bacterial or viral, or of grafted tissue, normally evokes an immunologic reaction, the ultimate effect of which is to destroy the invading organism, or foreign tissue, and thus preserve the integrity of the individual. A mechanism such as this, which undoubtedly evolved for reasons that should be evident from the examples cited, has obvious survival advantages. However, there is the concomitant disadvantage that vital organs, or skin, cannot be transplanted from another healthy individual because of the operation of this same phenomenon.

We have just mentioned that the mechanism for recognition of self is active during fetal development. It was predicted by Burnet, of Australia, and then later shown by Medawar, of England, that injection into the fetus of an inbred mouse of tissue suspension from another inbred line results in an adult that is then able to accept skin graft from the breed that provided the injected tissue. It appears that the introduction of substances into the developing animal, at a point in time prior to the establishment of the recognition mechanism, results in the foreign material then being regarded as self, rather than not-self.

It has furthermore been found that suitable injections into new-born animals, shortly after birth, will induce a similar state of immunologic tolerance for the particular antigen that is injected before this crucial moment. It is clear that the individual is still in a plastic state, even after birth, and that there is a decisive point in time before which tolerance will be induced, but after which the same treatment will induce immunity, which is sometimes manifest as a hastening of the normal reaction of rejection.

By analogy these phenomena suggest that the determinants of adult behavior are only in part genetically linked, and that other factors that influence perception and behavior are acquired, or modified, by events that produce their effects *in utero* or in early, or later, life. Such influences affect the somatoplasm of the individual and not the germ plasm, and hence are not transmissible by genetic inheritance. However, they do determine the characteristics and behavior of the individual during his lifetime and, acting as a kind of environmental influence, could affect others indirectly, including his own offspring and therefore succeeding generations.

Might there be any value, at least by analogy, for understanding animal and human behavior, in the processes involved in the phenomenon of induced immunologic tolerance? It is conceivable that reaction to life's experiences is of one kind if introduced before a certain decisive moment of development, and is quite different if introduced later. Is it conceivable that racial intolerance, as one example of intolerant behavior, may follow laws similar to those which govern the development of immunologic intolerance?

When we speak of immunologic tolerance or intolerance, or other kinds of intolerance, as in human behavior, we imply the existence of a memory mechanism; it is as if the organism subsequently responded with recognition and recollection, and with a power of discrimination in accordance with a pattern determined at the decisive moment in the developmental timetable. Thus the injection of an antigen at a time when immunologic tolerance can still be induced will result thereafter in reactions of acceptance, or tolerance, whereas the injection of the same antigen at a later time will result in antibody formation, and in intolerance, or rejection. Re-exposure to the antigen will result, accordingly, either in acceptance or in accelerated rejection, just as seems to occur in other learning experiences.

We need not go into further detail to illustrate the suggested analogy between the conditioned reflex in physiology and psychology and the "conditioned reflex" that seems to operate in the booster-type response to a subsequent exposure to an antigen in previously conditioned animals. However, it is of interest that quantitative immunologic studies have revealed that the intensity of the primary stimulus is a highly critical factor in determining the character of the response to a secondary or booster stimulus. Thus, if a useful analogy does exist between the conditioned reflex of psychology and physiology and the immunologic conditioning phenomenon, then patterns of possible interest in psychological phenomena could be observed through quantitative studies in immunology which might shed light upon questions that are, perhaps, not as readily perceivable and approachable in the psychological realm.

Still other immunologic phenomena are reminiscent of psychologic phenomena. One is referred to as immunologic paralysis, in which failure to respond to further injections of antigen can be induced if large doses are given at frequent intervals. This is not unlike failure of psychologic response, or of learning, under circumstances of excessive stimulation. Similarly, there can be failure of immunologic response through interference by too many, or by competing, antigens. Then there is the refractory period, after primary immunization, when a lapse of time is required before a booster response can be elicited, which perhaps corresponds to the period in the learning process required for assembling impressions and their fixation and preparation for recognition upon subsequent stimulation of memory by the same or other stimuli.

There is the familiar allergic reaction to certain exogenous antigens (i.e., introduced from without) and diseases of autoimmunization in which the individual forms antibodies to his own tissues. These are examples of exaggerated reactions, or of abnormalities in functioning, that may be said to be analogous to exaggerated psychological reactions to external events, or to the self-consuming effects of defense reactions, which instead of expressing themselves appropriately only against external threat, produce destructive effects upon the self.

Another familiar immunologic phenomenon is that referred to as passive as compared with active immunization. It is possible to induce a temporary effect of immunity by transferring antibodies from one host to another; but long-term immunizing effects can be induced only by the active participation of the host in developing his own antibodies as a consequence of his own interaction with the antigen. This phenomenon is not dissimilar to the effect observed in the individual who acts passively in response to what he is told but who has not, through engagement, learned in a way that would result in understanding and hence in the more durable effect of active experience. Thus it would appear that the process of "learning" in immunology, or in psychology, is something that involves active effort, and that what is learned is significant and effective in proportion to the effort expended. Thus a good antigen given in adequate dosage and on an appropriate schedule to a reactive individual will result in a substantial antibody response. The effect of this will persist for a long time. The analogy to the educational process needs no further amplification.

Another immunologic phenomenon with a seemingly analogous psychological counterpart is suggested by studies in germfree animals. It has been observed that animals reared in a germfree environment possess the immuno-cellular mechanisms for immunologic defense, but in a latent and essentially inactive or poorly developed state. For example, the lymphoid tissue is virtually nonexistent, and the gamma fraction of serum globulin, which has antibody properties, is present in very low concentrations or is not detectable. The introduction of infectious agents into "normal" animals activates lymphoid tissue to multiplication to form specific antibodies. However, the nonuse of this mechanism early in the life of the germ-free animal leaves it ill-prepared to meet life-threatening emergencies that may arise later in life. If the analogy were extended, this would be comparable to the inadequately trained mind—or to children reared under completely protected circumstances, resulting in individuals who can exist only in a protected environment and who are not prepared for the threats and challenges of life.

Carrying the analogy between immunologic and psychologic phenomena one step further, the mechanism by which psychologic learning takes place may conform to that suggested by the "selection" rather than the "instruction" theory of antibody formation—to use Lederberg's terms. This theory implies that antibody-forming cells possess the capacity to react to antigens selectively rather than that any antibody-forming cell can be instructed to form an antibody to any antigen. By analogy this would mean that there pre-exists in the CNS (central nervous system) a latent capacity for the development of the characteristics, or reaction patterns, that are later exhibited; the genetically determined pattern existing in the protoplasm is not expressed until impinged upon by circumstances in the environment in such a way as to develop the skills, thoughts, actions, and personality that eventually characterize each individual.

If psychologic and immunologic phenomena do represent different aspects of the mechanisms evolved with survival advantages, either for protection of the organism against adverse circumstances or for adaptation, might not some practical thoughts emerge from these imaginings? It is possible that, by analogy, knowledge of the development of the immunologic mechanism would be useful in understanding the development of psychological behavior and in preventing the development of pathologic or unhealthy states, reversible only with the greatest of difficulty or not at all.

Implicit in these analogies is the possibility that factors not now understood might, when better comprehended, be used to enhance man's adaptability and augment his effectiveness in coping with the large and rapid changes which he is contributing to an environment that has not heretofore taxed his capacity for adaptability to the extent that now prevails.

Instincts and the capacity for learning evolved, as did immunologic phenomena, as necessary requirements for coping with environmental influences encountered early in man's evolution, as well as in the lifetime of the individual. If in the evolutionary process man developed both an intuition and an intellect—which may be equated with instinctive reactions and with learned reactions, respectively—then the function and purpose served by intuition and intellect are as important to understand as any of the other biological traits of man.

Instinctive reactions and learned reactions operate to maintain the integrity of the individual in the psychological and social environment in which he finds himself. Through his intuitive sense and his intelligence the individual thus copes with life's experiences. While this is similar to the behavior of all living organisms, in the case of man the greater complexity of these functions makes him a special creature for study and understanding.

In the course of time man has so altered his environment, at a rate so far exceeding the development of changes in instinctive behavior, or intuitive reactions, that he is now the victim of stresses upon himself to the point where his intellect, which seems about to overpower him, must be invoked to help save him.

From all that has already been said it may be deduced that man is, fundamentally, a system of dualisms and that this is manifest in his function as a creative constructive individual as well as in other ways. The dualism in the intuitional and intellectual mechanism corresponds, in a sense, to the genetic and somatic mechanisms that are linked in simultaneous evolution. The essential unity of organism and environment, of genes and soma, and of intuition and intellect makes it clear that dissociation of one from the other leads to disintegration of the whole in ways that impair health and threaten life itself.

Man's constructive aim seems to be to bring about consonance between his outer environment and his inner self. His intellect needs to guide his intuitive sense as he strives to create a world closer to his heart's desire, as well as to control those forces in nature, such as agents of disease, that threaten his physical integrity. But man himself is also a harmful force in nature; he can impair his own personal integrity, and often acts against himself as if he himself were an agent of disease. When his physical or personal integrity is menaced, he defends himself against, or attacks, those whom he regards as threats.

Thus man needs to understand the relationship of his chronic feelings of threat, insecurity, intolerance, or tolerance to the forces operating within him and upon him in his struggle to survive and to maintain the integrity of his self.

4

Environment and Evolution

As previously stated, living things in general are an integral part of change in external factors. It is evident, too, that they must deal with forces arising from within. Both the forces from within and the outside factors change in the course of time. In human life, as the internal synthesizing systems develop and mature, changes in feelings and behavior occur. From time to time we sense new urges and desires, new forces that have arisen from within in reaction to the environment. These manifestations create additional problems or challenges, and require that we constantly cope with new situations. Thus we are exposed not only to new external factors and new internal forces, but also to new circumstances resulting from the interaction of both. Let us not only follow these ideas intellectually, but try to feel them as well.

From the moment of conception, and also at birth, each of us is essentially a "package of potential." The pattern of the potential within each of us is in constant emergence. We sense a desire to master our own destiny, and we wish to become increasingly responsible for ourselves. The influence of those on whom we are first dependent gradually diminishes and is eventually withdrawn, although with inevitable effects. In the same manner as those who were once responsible for us, we, in turn, become responsible for others, whose dependence upon us is transient.

As with all living things, our potential emerges under the influence of our environment. If we understand the relationship between the two, we may be able better to deal with the factors in the internal and external environments with which we are constantly confronted.

The individuality we are to express comes into existence at the moment of pairing of the chromosomes which contain the code of information that defines our "potential." At birth this individuality is recognizably different from that of any other human. From the moment of birth environment changes, and a long series of events begins to influence, subsequent development and the unfolding of our potential.

When we think of our potential in this way, we can visualize something which is not revealed until it is unlocked by long exposure to the conditions and circumstances of our lives—our environments. Environments differ in many ways, and the possibilities in us are educed by circumstances specific to their evocation. Before emergence our innate potential is unknowable; there is no way to perceive what will emerge, nor the specific factors needed to reveal as yet undisclosed potential.

Essentially, this is the situation in which we all find ourselves by virtue of our existence. At any point in our lives, we know what has happened in our development but not what will happen in the future. And yet we would like to know what to do, what role to play in determining our own future. In what way does chance, or luck, or other factors not under our direct control, operate?

The terms "environment" and "potential" are two abstractions, to which we shall attempt to give substance. In the sense in which we use the words here, "environment" implies a source of "information" and "potential" implies the capacity to "respond." For information to be effective, a reacting structure is needed. The operation of such phenomena in living systems is illustrated by the following example from the studies of molecular biologists:

Certain strains of bacillus coli are capable of producing an enzyme known as galactosidase. An enzyme is a specialized protein that possesses the property of digesting a specific substance, in this instance a sugar which is reduced to simpler elements by specific action at particular linkages in the sugar molecule. Some enzymes break complex molecules into their component elements, others synthesize simple elements into complex molecules. The particular enzyme to which we refer digests the sugar lactose, reducing it to galactose and glucose, and eventually these are reduced to carbon dioxide and water.

The capacity to digest lactose is not possessed by all bacteria. It is due to a gene which, in effect, is a unit of hereditary information that is expressed as an enzyme function. A bacterium that does not possess this gene will not have this digestive capacity. Thus the potential for digesting lactose is genetically determined.

Let us examine further the factors that determine the presence of the enzyme in the bacterial cell. We have said that lactose can be digested by a particular enzyme. Is this enzyme constantly present or is it merely the potential for producing the enzyme that is present? It appears that the enzyme is not always present but that the potential to produce it, seemingly as needed, does exist. How does this occur?

In a genetically competent bacterium, in which the enzyme is not present, the addition of lactose to a culture medium results in the production of the enzyme, which thereupon digests the lactose. Lactose acts to remove an inhibitor that normally keeps galactosidase production in check. It is as if the potential for producing the enzyme were in constant readiness, waiting only for the opportunity, or the need, to be expressed. Such an opportunity is afforded by lactose, which, by inactivating the inhibitor, behaves as if it were, in a sense, a stimulating or an evoking "environmental factor." The ultimate effect of the action of lactose, through inducing the formation of galactosidase, is to change the "environment" of the bacterium—i.e., to induce its own destruction.

This process describes an organism's reaction to an environmental agent in which the organism builds more of its own substance. Thus the bacterium is, essentially, a living machine that converts elements from its environment into its own substance. For so doing, remarkable mechanisms evolved for converting "biological potential" into "biological substance" from "environmental substance." This mechanism, however it first came into existence, has evolved along many different pathways —but in all instances the basic pattern is essentially the same. If we understand the pattern and its elements, then we can comprehend other, more complex relationships of the same basic character.

Another example with a suggestively similar pattern is antibody formation.

We noted earlier that an antibody is a protein formed in certain cells as a result of recovery from infection, or of vaccination, and is found in the bloodstream. It combines with the microbe that stimulates its formation and thereby contributes to recovery from, or prevents an attack of, the disease. Our interest here is in the process whereby antibody formation is induced.

An antibody is normally not produced unless there has been prior exposure to the antigen that evokes its formation. The capacity to form a particular antibody pre-exists, and the antigen—which is, in effect, an "antibody generator"—activates the antibody-forming mechanism. In comparison with the process of induced enzyme synthesis, the antigen may be regarded as an "environmental influence" acting upon the immunologic system, analogous to lactose, which as an "environmental influence" activates the enzyme-producing system of the bacterium.

Due to a genetic deficiency, some individuals are incapable of forming antibodies to particular antigens, in the same way as the bacterium referred to above may not be capable of producing the enzyme that digests lactose. If the potential for reacting to such "environmental substances" has survival value, it is through the operation of such mechanisms that "environmental factors" contribute to the form and design of living things and that the survivors, in turn, influence their environments.

To summarize: the presence of lactose in the bacterial culture, or of an antigen in a higher organism, has the effect of altering a genetically programmed state of equilibrium and inducing a reaction to destroy the provoking agent. The lactose or the antigen acts as a release-factor for already established genetically controlled programs and mechanisms which have survival value for the organism and are triggered by change.

Still other relationships between "environmental factors" and "organisms" illustrate the many different ways whereby living entities survive and maintain their identity. For example, certain bacteria and viruses persist inside a host by remaining innocuous, or even by being useful to the host. Among these are the useful intestinal bacteria which produce vitamins essential for life and the harmless bacteria in the throat that protect against harmful fungi or molds. The herpes simplex virus may be harmlessly present in nerve cells and only occasionally becomes active, causing the familiar "cold sore" on the lip or the more serious ulcer of the cornea that causes blindness. These bacteria or viruses are, in effect, "external environmental influences," although they are present in "the interior" of the host and are capable of affecting the host in various ways. Thus a normally external environmental factor when incorporated internally is capable of acting positively or negatively. This is in contrast to the negative effect of "pathogenic" microbes.

From this point of view, we look upon lactose as a food and upon an antigen as a stimulator of protective substances. Each may then be looked upon as a constructive environmental factor which evokes the organism's potential for sustenance or for defense against destruction.

I have tried to indicate the way in which "external environmental influences" and "biological potential" for survival are related, and to reveal that environmental influences may be potentially constructive as well as potentially destructive. I have noted differences among the so-called "environmental influences" which when assimilated act as food, as, for example, lactose, or antigens, which evoke self-protective responses, and the other organisms, such as bacteria or viruses, which when incorporated become internalized portions of the external environment and are potentially beneficial, or harmful, or exist without discernible effect.

It is clear from the discussion of induced immunologic tolerance in the previous chapter that there are decisive moments in development prior to which a given event will cause one effect and subsequent to which the same event will cause an opposite effect.

It was pointed out that the introduction of adult tissue from one line of inbred mouse into the fetus of a different inbred line resulted in the induction of tolerance on the part of the injected animal, as manifest by an acquired capacity to accept a skin graft. Conversely, animals so treated after birth rejected such grafts more rapidly than did untreated controls.

The point of this illustration is the recognition of the existence of decisive moments prior to which a given treatment produces one effect and after which it produces another effect. Critical periods thus exist not only in embryologic development but particularly in the maturation of the immunologic system; this is also seen in the development of the individual after birth as the result of early experiences in life and in education. In the course of early life, fixations occur at fixed critical periods as becoming ceases and existence begins.

Another example of this kind is a genetic defect in a biochemical process that influences mental development, which leads to mental retardation if a specific amino acid is not withheld prior to a particular critical period in development. The chemical involved is the amino acid phenylalanine. The condition is known clinically as phenylketonuria. It is revealed by the presence of phenylketone substances in the urine of an infant, which indicates an abnormality in phenylalanine metabolism. The presence of an excess of phenylalanine has an adverse effect on certain molecular events necessary for full mental development. If phenylalanine is withheld in the early weeks after birth, the infant will develop normally; if not, then a mentally defective child will result. This condition is caused primarily by a genetic defect in phenylalanine metabolism, which is manifest as an hereditary defect in intelligence. This, therefore, is another example of an effect attributable to the interplay between "genetic potential" and a kind of "environmental factor," which, in this instance, by being withheld, is capable of influencing a negative potential in a positive way.

These examples of purely biological phenomena are, by analogy, highly suggestive when we think about the processes of development in early life and in education.

With these analogies in mind, let us ask the question, "What is the potential of man and what is the nature of the environmental factors or forces needed for its fulfillment?" Answers may be found more readily if this question is kept in the foreground and if we learn to use our biologically rooted inner sense of direction.

If, from the beginning of life, opportunity exists for the active attainment of needs and gratification, the individual develops a sense of fulfillment and a basis for a system of values. If early life is characterized by easily or passively acquired gratification, or on the other hand by frustrated gratification, it is easy to see how defects in personality can develop at decisive moments in the long formative period after birth. This period, necessary for the development of the higher nervous system functions, is equivalent to the long prenatal period required for the development of the physical and functional systems necessary to life.

In the universe of living things environmental influences are not only external but also become incorporated internally. Thus the internal environment of living things contains elements which at earlier stages of evolution were part of the external environment. This is evident in the similar chemical composition of blood and of sea water which is due to the chance occurrence of the genetic capability and the survival of forms which possessed the capacity for incorporating the ingredients of sea water when useful for survival.

Thus the nature of a living organism derives, in part, from the nature of the internalized environment that it acquired in the course of evolution to fit its survival needs, and in accordance with a pre-existing pattern of potentiality established by that evolution. Similarly, in the unfolding of the unknown pattern in each of our lives, the incorporation of "environmental" or "experiential" factors, relating both to survival and to satisfactions, contributes to the development of the potential with which we are born. In addition, we have the desire to know about patterns and potentials of which we are not conscious. And through consciousness of our feelings and their meaning we can find these out. Through the experience of many opportunities, we begin to recognize how our capacities and capabilities can emerge. We are able to discover, in due course and within given limits, the opportunities that would draw out our particular potential. In this way we may find that we have many potentialities, and that there are many different ways in which these can be expressed.

We must remain aware of the need for matching our potential with favorable environmental influences; when such a match occurs, we must seize the opportunity if our development is to go forward. Too often it is the individual himself, rather than others, who constitutes the greatest limitation in his development and evolution. Those who desire to help us fulfill ourselves can be of the greatest help. They are friends who sense our needs, potentials, and patterns and who bring out the best in us. They serve as "environmental forces" to nurture and encourage the fulfillment of our potentialities.

It is through relationships such as these—whether between parent and child, teacher and student, lovers, peers, or between groups or nations—that the character of man can change, to become more constructive, more human, and more realistically related to a more complementary existence. Such relationships arrange themselves through reciprocal feelings and responses, which are expressed in individual behavior, whose character, under such circumstances, is likely to be more cooperative, more constructive, and more creative than otherwise.

The new and deeper knowledge of the workings of biological

systems that we now possess has suggested a way of looking at what is here broadly referred to as the "environment." By and large, environment has been looked upon as threatening and unfriendly. I would like to suggest another attitude toward the environment, considering it as a positive evolutionary force which affords opportunity for revealing the undisclosed potential that lies deep within man.

If environment is looked upon in this way, it is even possible to see something positive in what at first appears to be negative. Perhaps a more friendly attitude may thereby be developed and maintained toward the difficulties and adversities of living, in the course of man's perpetual quest for a better way of life, which seems constantly to renew and to correct itself. Since each day is a new beginning, we must examine objectively, as well as subjectively, each "adversity" for its positive value as we try to make the most of our potentialities and those of others.

An all too common attitude toward environment conceives of it as something to be conquered, something to be resented or overcome, rather than something to be incorporated for whatever values it may have for human growth, development and evolution. But environment is here looked upon not merely as if it were antagonistic but also as having value precisely because of its apparent opposition. This attitude presupposes that living things, including man, require "antagonism and adversity," or resistance, as part of the process of growth, development, and evolution. Selection in nature seems to consist of the survival of those forms best able to relate to environmental factors by utilizing what seems to be antagonistic for the advantage it may afford. To free human life of antagonism, or resistance, adversity, and therefore challenge, would be to deprive it of elements akin to food, which is part of the basic process essential to its fulfillment. I do not mean to imply that antagonism and adversity cannot be excessive, but rather that total conquest or elimination of adversity or resistance in the environment would also be excessive. Thus a balanced view of excess and insufficiency is needed in our attitude toward environmental and experiential influences in strengthening and fulfilling the potential for human life, recognizing that what is advantageous to some may be either without meaning or even disadvantageous to others.

This view suggests that the search for solutions to the problems of human life must be undertaken with the understanding that biological potential and environmental factors are complementary and not exclusive of one another. These forces are in constant interplay; where possible, a reasonable attitude toward the necessity for an environment of resistance and challenge rather than one of purity and sterility is more appropriate from the biological point of view.

5

Two Interrelated Value Systems—The Genetic and the Somatic

Biology deals, by and large, with the consequences of evolution, with the accumulated effects of past experiences, with mechanisms that earlier in evolutionary time were relatively simple and evolved slowly to the complexity we find in man. Medicine is concerned primarily with man's present health; it deals with the disorders of his body and his mind. As an object of study, human life is concerned largely with the problems of coping with uncertainties; it consists of man's expectations, hopes, and aspirations. It is difficult to separate biology from medicine, or to separate either from questions about human life; therefore, they are all of a piece.

The same forces of evolution which converged in the creation of man have now made possible, through man's work, the fission of the atom, the transformation of matter into energy, the creation of chemical elements not previously found on earth, the prolongation of the average life expectancy, and the control of the fate of other living species, as well as his own.

Through his capacity to examine himself, as well as the nature of things, man has discovered that he, too, is composed of atoms and of molecules. But this in itself does not explain how he functions. There is value in man's continued probing into the nature of things through deeper analysis leading to an understanding of himself for improving his effectiveness and sustaining his hopes.

As we have said, everything exists in an environment. For living things, environment must be regarded as if it were a part of the organism itself. As part of its substance, a living thing possesses an "internal" as well as an "external" environment.

Living organisms are dependent upon external factors for existence. From the simple observation that plant life requires sunlight, and that animal life feeds upon plant life, it follows that in the beginning life on earth depended upon light from the sun. Just as the earth itself was born from the sun, in the end life on earth will be extinguished by the extinction of the light from the sun.

At a certain point in time a molecule appeared that developed the capacity to copy itself. This was the moment when replication occurred in the precursors of what has now become the genetic DNA molecule. The simplest view would be that there appeared for the first time a "molecular machine" that could convert matter and energy into more of its own substance, according to the pattern of order then existing.

The molecular masses with the capacity for self-copying also possessed the capacity for change and the capacity for interacting with and organizing other environmental substances. Such newly organized elements, in turn, had an effect upon the molecules that produced them, and this interaction further accelerated change.

We have identified two basic interacting elements: (a) the replicating molecule and (b) the materials in the environment necessary for replication. In addition, replicating molecules require (c) a source of energy. The molecules which survived in evolution did so by virtue of an evolving means for utilizing energy derived from the sun. This source of energy provided a means for the first "living" molecules, and later living organisms, to evolve as rapidly as the capacity for conversion and storage of energy evolved. It would seem, therefore, that living things multiply at the expense of complementary environmental elements, when energy is available in a usable form.

The dependence of "matter," which possesses biological and evolutionary "potential," upon "environment" for raw material, and upon a source of "energy" for survival, for multiplication, and for change, is reflected in the character of successful evolution. This depended upon the development of the capacity for utilizing environmental elements and means for the conversion, storage, and use of solar energy.

An internal, self-controlled environment was made possible through the development of a cell membrane, which enclosed the elements created by and required by the evolving genetic material. Utilization of energy from the sun was facilitated by the development of a chlorophyll-like substance for conversion of solar energy into a biologically useful form. Thus the primordial lake which constituted the environment of the first precursors of DNA molecules was, in effect, reduced to the size of the contents of the cell membrane in which elements required for replication could be effectively concentrated.

Although usable sources of raw material and of energy could be stored inside the cell, replenishment through the semipermeable membrane was necessary for conversion into forms suitable for whatever functions were required. The cell as a whole was now the unit that expressed the biological and evolutionary potential of its genetic DNA; the external environment was outside the cell wall. By this development, an internal environment had been created.

Thus an increase in living material affects not only the environment of its own substance but the environment of other living substances as well. From this it would follow that, as time goes on, the "environmental" effect of living matter becomes greater and greater merely through the replication of genetic material. While replicating genetic material and environmental factors influence one another in ways directed in part by the genetic material itself and in part by environmental influences, chance plays a major role in determining the direction of change.

At this point in evolutionary time we see almost infinite complexity in the structure and function of living things; this is especially evident in man, who possesses an imagination which has been so effectively applied in the pursuit called science. The effects produced thereby are so complex that for greater comprehension we long for a measure of simplification. Although simplification will not eliminate the need to deal with the complexity of reality, it may furnish a point of view which can help in understanding complexity and, thereby, provide a conceptual frame within which it may be more readily tolerated. Ultimately, a simplified understanding will come through comprehension of the universe as a whole, including man.

The evolution of an internal environment has brought about the development of human intelligence, imagination, and ingenuity, qualities that resemble the "intelligence," "imagination," and "ingenuity" which nature has manifested in the changes that have led to man. The effect produced by man's own "internal milieu" upon his external environment, upon other living things, including humans, has, in many ways, become a greater source of disturbance for him than has the "external environment" of nature.

The products of man's imagination and his undisciplined appetites may have a boomerang effect which in due time may well overpower him. If his imagination becomes further uninhibited, his confidence more unlimited, and his conscience further obliterated, he faces a self-devised disaster. There is within him a sum of courage which has brought into existence all he has accomplished. But we might question whether he is taking the appropriate and requisite responsibility for his own future.

Let us assume, for purposes of discussion, a simple view of man and suggest means whereby he may then exercise control over himself. If the realization of man's potential has begun to exceed his physical, mental, and emotional capacities for coping, attention must be focused on his internal milieu to discover to what extent control may be possible. We may find, and therefore be forced to admit, that this is very limited indeed. However, the question must be asked deliberately if we are someday to have an answer.

Let us refer once more to the model we have used in the course of the discussion thus far. To recapitulate: The DNA molecule possesses an encoded potential that cannot be known until it is released through environmental influences. In effect, the potential in DNA is educed by environment. The problems with which the first DNA precursors were confronted were related entirely to survival, without change, even without growth and development—simple survival, which would permit the possibility of later change, growth, and development.

The precursors of the genetic DNA molecule have come a long way in "solving the problem" of "survival" for evolution. Living things that later evolved have, by definition, solved not only the problem of perpetuation and change, with innovations in various directions and at differing rates, depending upon accidental changes within the genetic molecule itself, but have also "solved the problem" of the chance effects resulting from environmental influences.

This occurred through the evolution of a dual system, the two parts of which are expressed in the genetic and the somatic components of a living organism. Although the somatic system is largely under the control of the genetic system, both exist in an inseparable, mutual relation of dependence. It is as if somatic structures had evolved to provide the means for replication of the genetic structure, analogous to the chicken and the egg, where each depends upon the other for existence.

If we consider the genetic and the somatic systems separately in terms of "biological potential," "biological energy," and "biological environment," then each has its own "potential," its own source of "energy," and its own "environment." Although both are independent, both are interdependently related.

Let us assume that early in evolutionary time the precursor of the genetic system appeared prior to the precursor of the somatic system, when endurance—survival—was of primary importance for later evolution. A second requisite for evolution was perpetuation through replication. To this the precursors of the somatic system made an important contribution. In the course of replication the evolutionary changes of greatest value were the development of mechanisms dealing with the environmental material required for replication and for energizing this process. As the genetic-somatic processes became more and more complex and interdependent, the predominant differences in the "value" of each merged, and living organisms became equally dependent upon the need of both the genetic and somatic systems for perpetuation and for expression of their biological and evolutionary potential.

The reason for making this distinction between the genetic and somatic processes, stressing the difference in purpose served by each at the time of their evolutionary appearance, is to emphasize their value-differences in biological evolution. It is probable that value-differences which existed early in evolution later became incorporated into the living organisms that subsequently evolved. Since man evolved from earlier forms of life, the problem of human values may be related to the system of biological values. What will be the subsequent evolutionary destiny of these two different primordial biological values (a) for existence now and (b) for perpetuation in the future? In what way are they evident and expressed in man?

Some men are concerned principally with what is of immediate value, while others are concerned with what will be of value in the future as well.

There seems to be a need to develop ways and means of safeguarding both the present and the future. The conflicts that arise within and among men are often due to conflicts of interest regarding present value as opposed to future value or present security as against future security. Often people are critical of one another because of differences in judgment as to the present consequences and the future implications of a position or an action. Such differences essentially reflect conflicting views of what each sees to be in his own interest, either immediate or future.

A new development in evolution may be the appearance in man of a more highly developed sense of the future, possibly contributing to increased anxiety, not only from immediate threats but from future threats, real or imagined. Threat for man is often a matter of interpretation, and reactions to threats frequently are related to previous experiences, whether acquired earlier and present now as automatic reactions with survival value (expressed as instincts), or reaction patterns which developed in the individual in the course of his struggle to express his physical or mental potential.

Although the mind and body of man are both somatic expressions of the germ plasm, each contains in a combined form the value-attributes which characterized the earliest forms of both the genetic and somatic systems. Thus survival and evolution have depended upon the combined and balanced presence of the value-attributes of both systems with respect to present and future needs; these two sets of value-attributes must, therefore, be present in more or less balanced combination in all spheres of human existence.

6

Purpose—A Biological Necessity

In the immediate sense, man is part of the universe of living things. Remotely, the substance of which man is composed was the substance of dead matter of the physical universe. We can understand the physical universe in terms of physics and chemistry, but these branches of knowledge alone do not provide sufficient insight for a full understanding of biologic systems; nor, for example, do physics and chemistry alone provide sufficient basis for understanding machines made by man.

Machines and biologic systems are analogous in that neither can be comprehended merely through a study of their parts. From the biologic viewpoint, parts can best be understood in relationship to the whole. Man, the individual, who is part of mankind, organized in societies, is analogous to individuals in subhuman species which are also organized socially. However, a single man differs from a single individual in species that do not form societies. The latter, in a way, are analogous to single molecules of a given variety in the physical universe. They are different from individuals in subhuman species that become socially arranged; man differs from long-fixed subhuman societies in that, culturally and socially, he is in rapid evolution. In a sense, societies are like an organism with many parts, each of which, and together with all, constitutes the organism. Each part, or unit, is potentially of equal importance for the work or fate of the organism as a whole.

Because it is basic to our further discussion, I want to underline the idea that societies—and mankind—consist of individuals with different kinds and degrees of purpose. In this way we can identify man's relationship to man. My aim is to emphasize the appropriateness of speaking of purpose in biology in general as well as in relation to man. I hasten to add that I think it entirely unnecessary to consider purpose in order to understand the phenomena of physics and chemistry. Living systems require different considerations than nonliving systems; the idea of purpose in living systems is not just relevant; it is essential. To explain what we mean by purpose, we must first develop several general ideas from biologic experience.

When molecules became organized into cells, living machines were formed which possessed structural elements in functionally meaningful relationships. The new possibilities of protoplasm were expressed successively through the augmentation of potentialities educed by previous circumstances.

"Knowledge" in the sense of "human knowledge" was not necessary for biological "inventions" developed in the course of evolution. Such new "knowledge," in the "nonconscious" sense, grew step by step as the capacity of biologic systems for nonconscious "learning" seemed to increase.

It should be evident that morphological and functional development, or "learning" in the phylogenetic and ontogenetic sense, was a process which merely exposed protoplasmic (i.e., genetic and somatic) possibilities that pre-existed. Such possibilities came into evidence because they were there. Some outside force—let us call it environmental—provided the suitable preconditions to the expression of this potential.

It is as if the environment exerted a force on the plastic organism which brings forth a potential. If a potential, whether in the germ plasm or in the somatoplasm, does not exist, then the environmental force is without effect; if a potential does exist, then the environmental force educes an effect.

Thus we may say that whatever change may have occurred was inherent in the protoplasm. Although nonconscious, the organism had no "choice" but to react in this way to the prevailing circumstances. The structural and functional changes, whether germ plasmic or somatoplasmic in nature, then became a part of the structure and function of the organism and of its capacity to react further.

It may be concluded that it is in the nature of the organism to be oriented for the change that occurs. The intrinsic nature of the organism influences the range and direction of change that can occur; the change is then added to others, all of which together seem to be "causes" toward which the developing organism is drawn. The word "cause," in this context, has the philosophic meaning of "end or purpose for which a thing is done or produced"—in the sense that we say "we work for a cause." The "cause" to which the organism is drawn, therefore, is contained in the germ plasm of the species and of the individual.

Thus it would seem that "causes" are present in the germ plasm—in the sense that the germ plasm contains the "formula" for survival or the capacity to react in other ways, ways that are not necessarily of immediate survival value. Thus "cause," "purpose," and "goal" are regarded as potentially present in man's protoplasmic substance, and as alterable from time to time within limits set by genetic factors.

Each developmental or evolutionary step occurs as a result of chance "genetic" changes in the organism and of its "somatic" responsiveness to environmental influences, which might be thought of as fields of force. This is not the same as saying that the evolutionary, or developmental, change, or the step, was for the purpose of achieving a particular goal, as would be the "explanation" in teleologic terms; rather, the effect or purpose served is, in a sense, the "cause" of the change required for survival and of the chance protoplasmic changes and responses.

Thus the development and evolution of living organisms is actively influenced by "environment." Although we may speak of the evolution of the species in the phylogenetic sense, or the development of the organism in the ontogenetic sense, both are the result of the operation of several pressures or fields of force.

It might logically follow that without "cause" or "purpose," in this sense, living organisms would not have evolved as they have, and it might follow that without "cause" or "purpose," human life, in its present form and design, would not exist. It seems, therefore, that "purpose" is an essential element of biological systems, and, empirically, "purpose" does seem to be vitally important for man—not only for man the individual but for human societies as well. Might this not also explain why there is tension and anxiety in relation to human purpose and human goals as well as to national purpose and national goals?

Societies and cultures create fields of force for change—hence for evolution. In the same sense, an individual creates a field of force that acts upon himself and contributes to the fields of force that act upon others. If we then speak of "causes" and of the effects induced thereby, we observe a directional influence upon human potential induced by environment—whether the environment is physical, social, or individual. It would appear, therefore, that "goal" and "purpose" are part of living systems. By definition, a living system does not exist, in the sense of being alive, without purpose, even if the purpose is merely that of staying alive.

Apart from the many different factors and many different sets of needs that have entered into the design of living organisms, they are at least programmed for survival. It should be no surprise, therefore, to find man conforming to other living systems in this respect. But for man there is a difference between survival and "living."

For man, his cultural evolution may be said to have begun when he could "select" options—when decisions were not all established in biologically automated systems within his protoplasm, when he became responsible for his choices, his decisions, and his acts. Because of this it seems as if each individual constituted a new and different "variety" among other humans and, for his own "survival," required the discovery and exploitation of his own attributes and strengths. Each individual was in some ways a new mutant or new variety, in the taxonomic sense, and existed in competition with other, similar varieties according to a competitive-exclusion principle. When societies first formed, they were also competitive until they developed alliances, much as do individuals—in healthy family groups or communities.

Each new human individual and each new generation possesses a capacity to respond to "callings" which are, in effect, "purposes" and "goals." New methods, materials, and institutions are devised to attain such ends. The long postnatal period is a time of preparation for adulthood. This is the time when the responsibility for decisions for self and for others begins to weigh heavily. How are we to prepare for this function of maturity?

We can do only some of the many things that we seem to need to do. We cannot satisfy every urge. On the other hand, there are times when we have a sense of futility because there seems to be nothing for us to do. The absence of purpose leads to a sense of nothingness, or emptiness—to a feeling of "want." This is an uncomfortable and, at times, an unhealthy state, the cause of which needs to be identified. Is it because there are no challenges? Is it because we are not sensitive to those that exist? Is it that we are not interested? It is conceivable that we are not conscious of new challenges and that we may be overly concerned with preoccupations that no longer satisfy us. What more is there for individuals to do? What more is there for a school or a profession, or for practitioners of an art or a science or a service, to do? What more is there for a community or a nation to do? What more is there for mankind to do?

Whatever the answers to these questions, we acknowledge the existence in each of us of different interests and desires which have to be satisfied. They may be intellectual, aesthetic, social, or personal. The exhortation to "know thyself" is based on a real need—we might call it a biological need—for an awareness of the nature of the special interests and desires which are in each of us. This awareness creates a demand for their development to the extent of our ability to develop them under the circumstances that prevail. "What is there in us that can be cultivated to bring satisfaction to ourselves—and in so doing become creative, contributing members of society?" When the answer to this question is not clear, then we have not yet reached a point of sufficient understanding for commitment. For each pattern to be expressed, a source of power is essential, and that source lies within each of us.

Society cannot assume, and should not be expected to assume, exclusive responsibility for identifying, nurturing, and encouraging abilities and talents. Individuals and society must share in the establishment of value through the approval of talents and orientations which give purpose or evolutionary direction to man, leading to still newer and higher purposes, revealing still further potentiality than has already been expressed. There are two independent cause-producing centers—the individual and society. Each creates goals and purposes, and ideally each should be devoted to the purposes of the other. Each could then serve itself best by best serving the purpose of the other.

This is both a personal and a social matter. It is important that we know it both for ourselves and for others. It would be appropriate in the early years of life to establish confidence in the self, permitting freedom to come to terms with one's self in these respects. The attitude toward life that might be expected to be developed would then be not "What's in it for me?" but rather "What can I do best and how?"

Our present state of advancement and unification of knowledge is such that it can reasonably be said that the generation now entering adulthood is the first to be in a position to draw upon sources of knowledge and inspiration to build a philosophy based on the operation in man of biological laws. When understood, these laws will provide a basis for the better use of man's power of self-development for ends which may be individually and collectively satisfying, although many problems will no doubt continue to confront us.

7

Responsibility

Although analogies can be seen between man and other forms of life, in respect to the various phenomena discussed in this book, man as a whole is different in very important ways from other living organisms or systems with which analogies can be drawn. For example, in forms of life other than man behavior analogous to "responsible" behavior is genetically programmed and essentially fixed. Such behavior seems to have developed as part of the accumulation of attributes valuable for survival. Individuals of the species not programmed to react "automatically" did not survive. Thus, in the course of evolution, "responsible" behavior tended to persist and to evolve with increasing complexity and subtlety, until in man it became a conscious "sense of responsibility."

As we think about "responsibility" in these terms, we recognize how basic it is for relationships in living systems generally. As an analogy, to take a nonhuman example, we sometimes refer to a cell as a complex society of molecules. For the order in living nature to be expressed, functional relationships with a high degree of "reliability" must exist in each subunit. Each of the component parts of a cell must "perform" its function "reliably," i.e., with what we might call "a sense of responsibility" not to itself alone but to the group of molecules, or to the larger society of molecules, of which it is a part. It is not difficult to understand the "disorder" that can occur when one of the component parts of a cell "behaves" in a way which, in human terms, might be interpreted as "irresponsible."

It is tautological to say that survival of a species, or of an organism, "depends" upon the success with which the problems of survival are managed. Nevertheless, it permits us to convey the idea that for an organism to be able to cope with the problems of survival, systems of molecules, cells, and organs must each behave "responsibly"—as if life "depended" upon the function and action of each of the component parts of the whole.

A familiar example of this is the relationship between a mother and her young; here "a sense of responsibility" on the part of the mother is critically important. When maternal responsibility is absent, as is sometimes seen under certain natural circumstances, and as has been induced experimentally in young monkeys, the young is neglected and often will not survive. It is not difficult to see that a species would not survive without a "built-in" sense of responsibility both for self and for others of the species.

Modern biologists speak of mechanisms of regulation and control as they apply to given functions, whether cell division, protein synthesis, or behavior. Speaking by analogy, the natural mechanism by which "responsibility" for a given function is regulated and controlled in the individual and in society is determined by an exercise of volition and choice expressive of differences among individuals with respect to their degree of self-concern as well as their concern for others.

Another way of expressing this idea is to say that each individual is programmed genetically to look after itself and after certain others of the species. In man the exercise of responsibility in this sense begins with the first cry at birth and continues to develop thereafter in all except those who are defective or deformed in this respect. The response of the mother to the infant is important at birth, and thereafter members of society usually respond to the needs of the young of the species if they are neglected, abused, or misled. This response is manifested by members of the family, by the public at large, and by government.

In man it is not very "debatable" as to whether infants should be allowed to struggle alone for survival, or be neglected or abused. However, "debate" does ensue when it becomes necessary to define the relative degree of responsibility to be assumed by the individual, by the public, or by the government for those members of society who are dependent upon others by virtue of handicap, age, social or economic deprivation.

In insect societies (bees or ants, for example) it is unlikely that such "debates" ensue. This is not to be construed as commending the organization of bee or ant societies as either a possible or a desirable prototype for human society, but rather to emphasize the obvious "natural" difference existing between societies of men and societies of insects.

Among insects, "judgments" or "choices" have been made by the process of natural selection in the course of biological evolution in terms of what is best for the survival of the species. Among men, some judgments are made that seem to be primarily of value not for the species but rather for the individual, and each individual retains an opinion about and some measure of control over his duty to himself and to others, and as to how much responsibility the public at large and the government should be expected to assume for him, and how much responsibility he should assume for others. Similarly, groups within the public decide in terms of what is in their best interest, while the government has a value system in which individual self-interest and public interest are in close competition.

To further illustrate the problem of responsibility in this sense, there is the generally available evidence which shows the association between cigarette smoking and the frequency of occurrence of certain disease states. The evidence suggests that if hazardous cigarettes did not exist there would follow a considerable reduction in disease and in premature death due to coronary artery disease and to cancer of the lung (and of other parts of the respiratory tract), and to cancer of the bladder. If a vaccine or a pill were devised that could control or postpone these diseases, there is little doubt that such procedures would eventually be widely applied. However, when what is involved is a voluntary change in a way of life, a change in smoking habits, for example, then we see the difficulty. Such difficulties arise when the individual is required to take "responsibility for himself"—to take action on his own behalf and, especially, "to give up something" that is personally meaningful to him, to which he may even have become addicted. The same may be said with respect to unhealthy dietary habits, to lack of physical exercise, or to changes required to develop a way of life involving less continuous stress, or to giving up potentially habit-forming drugs of various kinds, all of which have important bearing on the health and well-being of man.

Earlier I touched upon the social value of a sense of "responsibility for others," particularly for the young, at a time in life when such relationships are vitally necessary. I have just touched upon the individual and personal value of "responsibility for self." An example of "public responsibility" at a time when such a need existed was the formation in the United States in the 1930's of the National Foundation for Infantile Paralysis by a group of private citizens, to lead the public to take responsibility for a matter of importance to the public at large. The consequences of this action, through the work of the March of Dimes, is now clear. Later, the government assumed more and more responsibility for the support of biological and medical research and has come to play a greater and greater role in advancing knowledge that will improve the health of the nation and of man generally.

The future of man requires the definition of values and of purposes to be served and of the role that responsibility plays, not by chance, not by moral demand, but by "necessity" for survival. Our choice of values reflects an imbalance in men's thinking between interest in self and interest in others. This, in turn, reveals different attitudes toward values in human life.

Each new generation comes upon the scene with a new view and a new desire to assume its share of responsibility, since responsibility for self and for others is the expression of an innate biological necessity. We see evidence that the exercise of responsibility contributes to the health and well-being of individual man. We see this as a basic need which, if taken away, deprives him of his means for fulfilling himself, for becoming "greater" than his undeveloped, genetically determined, automatic reaction patterns would permit. Man, in this respect, differs from bees and ants and other animals. He has the ability to learn. He has already transformed the face of the earth by the exercise of his creative capacity. Now there is a need for him to reveal and to express his sense of responsibility for his species as well as for himself. Man possesses the capacity for many responsibilities, which each individual must identify and develop for himself.

The conflict within the mind of man is evident in our youth. Some respond with constructive rebellion, while others are destructive in their revolt. We are tempted to ask what would happen if representatives of the young were included respectfully, and without condescension, in the discussion and planning for the future in order that they, too, might share such responsibility. There is no more sobering effect upon those who seek control and power than that of sharing responsibility for the consequences which power implies.

The basic health of each new generation must be trusted to reject what is harmful and to recognize and correct the errors of the generations that have gone before. This has been the process that has guided man unconsciously since his beginning. Now he must pursue it consciously.

By acting responsibly toward self and toward others, a hierarchy of responsibility develops and soon becomes definable. Without order of this kind, man and society become utterly confused and lost. It is in this regard that responsibility is—even in human terms—a biological necessity.

"Responsible behavior," as has been exemplified, does contribute to the health and well-being of man individually, but may have an unhealthy effect if the health and well-being of both the individual and society, or the species, are not considered. When responsibility for the individual and the species is separated, disease and disorder result—between individuals, between groups, and between nations. The natural order cannot be violated without penalty. The rules adopted by man as a guide to obeying the natural laws must be realistically designed to fit the character of man, to which they apply. This is our responsibility and our challenge.

Can we make order out of chaos by any means other than by attempting to develop in the young a *healthy* sense of the order in nature and of responsibility to nature so that this sense may grow and mature over the years? The hope is that as experience accumulates in the course of the early years of life, the child, the adolescent, and the youth will develop a sense of responsibility for self and for others which will come into a reasonable balance as the early struggle for identity is won. In this way, each individual would contribute also to the development of a healthy sense of identity and responsibility in others.

With a healthy sense of one's self, a healthy sense of responsibility, and a purpose—knowing what one can do, what one needs to do, and what one wants to do—it becomes possible to make a commitment to a cause and to develop relationships in which each individual is a meaningful part of the whole, namely, the family, the group, and mankind.

We cannot blame those who have not yet lived long enough for not understanding the importance of lessons learned only in the course of living. Nor can we blame those who have already lived a good part of their lives for not understanding the nature of the confusion and the dilemma in the minds of those who are as yet inexperienced and uncommitted in the world as it is at the present time. The young often react negatively to those who coerce them into a conformity of behavior with which they cannot identify and which they cannot understand or believe.

It is necessary to develop a means of communication between the generations to help the young individual find himself even in an intolerant and unfriendly world. He tends to protect and defend himself until he realizes that there are others like him, of all ages, who believe as he does and with whom he may join. When he recognizes this, his self-esteem, his self-confidence, and his sense of responsibility to others are enlarged. When he successfully extends his sense of responsibility to others, he develops a sense of satisfaction, and an increasing desire to relate to others through whatever form of expression is natural to him, whether it is art, science, medicine, business, personal relationships, law or politics. This has been the basis of the success of man's evolutionary pattern thus far. Where man goes from here depends on how wisely he consciously assumes and discharges his responsibility as the "Trustee of his own Evolution" that he has become.

Many other characteristics and attributes of man, and many other aspects of his behavior, need to be viewed in a new light so that man may develop a deeper and more realistic view of himself. It seems possible now for man to view himself from a new vantage point, one built upon the contributions to general knowledge that have come from biological insights and understanding. It seems now that the epistemologic contributions that have come from biologists may be able to provide a basis for a more realistic philosophy that will lead man to see himself as he is—a mixture of good and evil—and not as he would like to imagine himself to be—all good or, as he sometimes conducts himself, all evil.

8

Awareness of Order

As has already been noted, there is recorded in the chromosomes of each of us a compilation of information accumulated in the course of evolutionary time—a set of specifications and operating instructions. This set of instructions and the systems for their translation into the mechanisms that keep individuals alive represent one of the most remarkable of "biological inventions." This "invention" is complex, and in it is contained one of man's most remarkable features—his awareness of himself. With this awareness he can also recognize the existence of patterns of order through which, so to speak, matter becomes "conscious" of itself.

This recognition gives rise to a number of interesting questions. What is the nature of order in living things, and what can be done consciously and therefore deliberately about the order within man and the responsibility for its development? This is a difficult question to answer with any degree of certainty.

We cannot alter the fact that each of us is unique. The combination of circumstances that gave rise to each was different and can never again be duplicated. Therefore, we must expect that in each of us there will be differences in desire as well as determination, discipline, and sense of responsibility, all attributes of man which make him an animal concerned with the future—a concern for more than just the perpetuation of the species.

A concern about the future implies a desire to know about the past, in part because it had an orienting influence upon what was once the future, though it is now read as the past. An awareness of the interrelatedness of the past, present, and future cannot be avoided. Nor can we avoid recognition of the distinguishable patterns that have emerged in the course of time. The life of each of us recapitulates an unfolding of the evolutionary history of man.

In the course of our lifetime we are responsible for performing in a way that will bring satisfaction to ourselves, and to others as well; others have the same kind of responsibility. Thus the responsibility for each and for all is shared. To the extent that it is shared, the greater is the awareness of ourselves in relation to others; the clearer also will be the awareness of order in ourselves as in others; and the greater will be the awareness of order in the group of which we are a part and in groups of which others are a part.

This awareness of order is accompanied by an awareness of the existence of value judgments. Value judgments are in part decisions relating to the self, and to the system of desires that exists within the self. In terms of self-development, this requires an understanding of how value judgments are constructed so that one can exercise control over one's own development. Our value judgments reflect our attitude; these in turn act as a kind of semipermeable membrane that selects and chooses. It is in respect to our value judgments that we come face to face with conflicts within the individual, among individuals, and among groups, and come face to face with the question of "good and evil."

Another accompaniment of awareness and of judgment is discovery. This process is endless since there will always be more to know than we can know at any one time. Not only do we learn things we have never known before, but we are able to see things differently and come to judgments different from those we may have previously made. We can also understand how others could do the same and how necessary it is to understand and accept that different patterns of order can exist at different times, in different people, and even simultaneously.

I shall not now develop the mental picture of how this comes about, lest, by introducing another idea, I fail to make the principal point. I want, primarily, to establish firmly the idea of our uniqueness and of the need to look inside as well as outside ourselves for direction and guidance.

We are all aware of order in the physical universe. We are also aware of change in the physical universe. This combination of order and change affords the opportunity for evolution. We are also familiar with evolution and, therefore, with order and change in the universe of living things. We would like now to emphasize the importance to each of us individually of the discovery and recognition of *the* pattern of order that exists within *each* of us—not only the pattern of order that is subject to change, but the pattern of order that influences change.

The discovery of patterns comes about in an interesting way. We recognize when something satisfies or dissatisfies. When this happens we sense it, we think about it, and then we know. Essentially, we cannot know without thinking. Therefore, sensing and thinking are both necessary for knowing. Thinking without sensing and sensing without thinking may be said to be wasteful if not used to lead beyond the moment. Sensing together with thought can lead to action, and this requires the operation of judgment.

There is a need to be aware of this deeply—to be aware of this deep in one's intuition. It is not enough to acknowledge intellectually the existence of differences between people, and the desirable effect of developing one's own talents in relation to the talents of others. For some the source of inspiration and judgment for discovery is within. For some it comes from without. In each instance we shall find ourselves desirous of knowing more and perhaps even of doing something about what we discover. Pictures develop in the minds of some of us; the minds of others are stimulated, and the effect of this is to enhance the development of the potential of all; this is the value of and the basis for the aggregation that occurs among men. Each aggregate when it occurs has a purpose, and the purpose is often not apparent until the effects of the aggregation are seen or felt.

When one thinks about the behavior of people, too often one sees evidence of a lack of deep awareness of what seems to us so obvious.

And yet there must be reasons for the seeming failure to acknowledge the existence of "constructive differences" among people. Perhaps this is because differences seem threatening and are potentially uprooting. Such an attitude sometimes results in withdrawal and isolation. If this withdrawal into one's self is for the development of a deeper understanding of nature or of man in nature, the effect of such withdrawal and communion with self, if communicated to others, will have the beneficial effect of participation. Through the collective mind benefits of thought and feeling can be transmitted to the minds of all who are able to enter into the discourse.

The effect of the development of the potential power of the human brain is to increase its potential for still further development. But here we are confronted by the question of value judgments because authority in any form may be used for good or for evil. It is about this that all men must be concerned.

We are confronted with the fact that the potential that exists within each of us depends upon the existence of an opportunity for its expression. Some of our most anxious moments are due to this dependence and this uncertainty. Often we should be grateful for our disappointments, since they can turn out to be our moments of greatest good fortune. Not infrequently disappointment is followed by an opportunity to bring out an unknown talent which could not have been realized without such disappointment. It has been said that there are two kinds of tragedies in life: those attributable to not getting what we want and the others to getting what we want.

The value of man's potential to adapt lies in his ability to convert what appears to be adversity into something of positive value.

The recognition that opportunity can come in many forms, and that there is more than one way in which the pattern that exists inside us can be drawn out, may relieve the feeling of anxiety that occurs the moment after disappointment is first felt. Awareness of these patterns can define the order of each day, and can affect the nature of the thoughts and feelings that occur within us and within others at any given time.

The greatest wealth a society can have lies in the opportunity that it affords its people, at all ages, to express the potential that exists in them so that they may be encouraged to develop primarily for their own sake.

Can it be said that one purpose in our lives is to express what awaits inside us to be called forth, and that the consequence of so doing is to further increase the changes and effects of opportunity? The increased opportunity may be for understanding and for action, and the value of this is measured by the quantity of constructive changes for the self and for others.

A consideration of the awareness of order has led to the recognition that the unfolding of orderly patterns, when opportunity exists for promising potential to be expressed, is an endless process, one of constant discovery and of new realization as patterns increase in complexity.

These processes are without end, and our interrelatedness to others in the groups of which we are a part, and of the groups themselves, reflects an endless pattern which appears to mirror the orderliness of a process rather than an end that can be defined. If the end can be stated for each of us individually, or for all of us collectively, it is to stay alive. If wishes could also be goals, to help others, who have the same desire, we must do so without compromising the existence of others. Whether this possibility lies within the nature of man is a question to which we have long addressed ourselves—especially when we are confronted with the contradiction of war. In spite of the destructive forces in man, there are also constructive forces which are part of the natural order.

Contradictions will never disappear. However, the nature of

order in living systems, and in man in particular, will in due course be better understood. We must increase our awareness of the order that exists within ourselves individually, and draw attention to the existence of order in others, which together make up the collective order of man.

Some of us, during our lifetimes, may become concerned with questions such as these. All of us will acquire greater experience in understanding ourselves and others, and see the varied patterns among others that lie beyond the limits of our day-to-day acquaintance. Some of us learn, firsthand, the nature of the difficult and complex problems among those who still struggle for survival, where the order in life is simply a struggle for sustenance, in contrast to the extraordinary opportunities for self-development available to those who are more fortunate.

It would be deplorable if, in the course of our experiences, we should develop the feeling that there is less and less opportunity, rather than that opportunity is always present. Where it is restricted or limited, unjustly, selfishly, or unreasonably, restriction must be overcome. It is in the nature of man to be free to raise the perfection of his performance so that he may develop constructive conduct for coping with threats to his existence and with factors that limit his development.

9

Change, Chance, Choice, and Challenge

"Change" is a word the meaning of which we take for granted—a word which is our language symbol for something that we simultaneously encourage and resist. Without change neither a living being nor evolution would have come into existence. Change demands adaptation; without change adaptation would be unnecessary and evolution would cease.

Not only is change reflected in the characteristics of living things, but no form of life exists without resistance or without some form of opposition. As we think this way, we recognize the existence of three other phenomena in human life—chance, choice, and challenge.

Before we discuss chance in human life, let us examine the operation of chance in other forms of life. Life could proceed in its evolutionary course only by virtue of the existence of the "opportunity," afforded by chance, to express the potential that existed in the earliest and simplest molecular configurations when life first began, and that at each successive moment in time is capable of becoming something different. The nature of evolutionary changes that have occurred, and that continue to occur, was influenced by chance. If the factors essential for survival had not "chanced" to occur, it is obvious that evolution would have taken a different direction or would not have continued at all.

The *presence* of "chance," or of "opportunity," in this sense, is a liberating and creative influence, while the *absence* of "chance," or of "opportunity," is a limiting or destroying influence. Living things possess a force capable of overcoming resistance to change, as the forms and the conditions of life change in time and with circumstances. "Change" is universal, and "chance" plays a major role in the changes that occur in each of the many small universes of which the universe at large is composed.

By implying the existence of many small but interrelated universes, including the nonliving and the living, we infer the existence of different constellations of matter in each of which chance has played a role of importance in their related courses of evolution. "Chance" is physically expressed in circumstances. It requires for its operation the existence of a material substance upon which it can play. It is clear that there must be complementarity between circumstances and the material that possesses the potential for change.

Under the circumstances that prevailed early in the evolution of living things, primordial living matter interacted with certain factors and not with others. "Choice" was dictated by the affinities that were the properties of early living matter, and could not have been known to have existed until exposed by chance to appropriate circumstances.

These are obvious and self-evident truths that apply as well to molecules as to man. What does it mean to each of us, in our respective lives, to recognize the effect of chance on the molecules that preceded us, and to recognize the continued effect of chance upon the man into which these molecules evolved?

At the time of birth we have no way of knowing what the child will be like when fully grown. We have come to recognize, even though chance plays a prominent role, that in an atmosphere of freedom and with a range of choice, the potential within the individual will express itself in its own unique way. We have also come to recognize that, for the exercise of choice by man, there must exist a basis for judgment. In this respect molecules have an easier problem than do men, since in man conscious choice implies an awareness or foreknowledge of consequences, and carries with it the implication of responsibility and risk. In the earliest forms of life foreknowledge could not have existed; correctness was judged by survival; responsibility had not yet been born.

A characteristic of all living things is the insistence to live; it may be said that living things exist by virtue of this insistence. The molecules of which living things are composed are created by the living substance itself, seemingly on demand, to serve a purpose connected with life. The substance of life may be said to be created and to exist by virtue of this demand. Thus living things both "insist" on living and "demand" mechanisms to serve this insistence. This was as true of primordial living matter as it is of man.

To apply this analogy to a higher order of complexity, we observe the way in which "insistence" by societies, through the purposes they profess, places "demands" for and upon the individuals of which they are composed. We observe the existence of a relationship between the individual and the group of which he is a part. If it were not for "relationships," societies would not exist, nor would organisms or cells, or the molecules of which cells and organisms are comprised. It would follow that if it were not for "relationships," man would not exist. Thus "relationships" are inextricably a characteristic of living substances.

We see the continuum from molecules to man, and we see evidences of the operation of forces which increase in complexity and intensity with the ultimate emergence in man of "will" and of "choice." It is as if the life force itself, whatever it may be, had evolved in time in a way that could not have been predicted prior to the emergence of the many forms in which it exists. This would include "man's will," which may be looked upon as an expression of the evolutionary development of the "insistence" of living things to survive. We shall not try to analyze further "man's will," or the "insistence" of living things to survive, but shall merely accept its existence and try to understand the forms that it has taken and the effect this has with respect to human choice.

The will of man, and his "desire" to live, we now see, is analogous to the affinities of chemical substances. It is also analogous to the needs and requirements of all living things for growth, development, and evolution for survival. The choices man tends to make are the result, in part, of the process of natural evolutionary testing and, in part, of the testing of the particular culture or society of which he is a member. Choices imprinted in protoplasm are part of the phenomenon called "instincts" or "needs."

Choice implies preference. Although molecules must also "decide," and the consequence of a "choice" is one effect or another, for man the problem is exceedingly more complex than for molecules. While this does not tell us anything that we don't already know, it may tell us that there is no escape from choice, whether the need is to decide "to do," "not to do," or "how much to do." Even this idea has been expressed many times before, and far more eloquently; but perhaps it may not have been related to models and systems familiar to the natural scientist, having been previously considered only by the philosophers and poets.

While it would appear that "choice," in the nonconscious as well as the conscious sense, is as much a part of life as are "change" and "chance," the "choices" confronting man, by their nature, and as they relate to his future, are such as to pose for him more than it may be reasonable to expect of him. The question before him is whether he can attain a state of understanding and control over the forces of which he himself is composed, through use of the power necessary for so doing, yet still remain in harmony with the flow of evolution. Must evolution, as expressed in man, proceed with such violence as to lead to disaster—if not for the entire human race, then certainly for the many who fall victim to choices which our experience, knowledge, and understanding judge to be "evil"? By "evil" I here mean the destruction of another at whose expense the destroyer survives, or otherwise benefits.

It would seem from all this that in the course of evolution

from molecules to man there has occurred a shift in relative importance from chance to choice. Man's life is not determined by chance alone; neither is choice always possible. But there does exist the challenge for man to create a world closer to one which men the world over desire.

By virtue of the "insistence of life to survive," a challenge is posed for all living things to invent mechanisms for increasing the probability of survival. Not until the advent of man has this insistence on survival become a *conscious, controllable* process, possessing a *voluntary* component, even to the point where man has developed a means for guaranteeing his own survival against a human enemy, the efficiency of which, however, assures the destruction not only of the enemy but of himself as well.

Circumstances evoking such thoughts are sufficiently prevalent to dominate the thinking and feeling of all who are aware of the world around them. In fact, a large part of the world's population and the world's resources is occupied in activities arising from this state of mind.

Can war-making in man be eliminated? This is one of the many questions and challenges still confronting man. It is one in which man has the power of choice. The challenge confronting man is to determine whether or not he is able to exercise the choices that will solve this problem, clearly his most important disorder—one that may even be thought of as a self-induced disease.

Before the methods and ways of thought of science were developed, many of the effects man observed in nature and among men were attributed to the activity of supernatural forces expressing their displeasure. Man himself expressed displeasure in ways leading to such conclusions; he therefore was describing his own nature when he attributed to the gods actions for which he could see no external cause. If man attributes to outside forces, or to others, causes for which he himself may well share responsibility, then the challenge confronting man is twofold—that of understanding and that of exercising control. One cannot expect man to act reasonably upon something that he does not understand. In the past, for example, when man became aware that many of his diseases were caused by outside agents, such as microbes, poisons, nutritional deficiencies, excesses of radiation, or by machines and structures created by him, he has in most instances acted appropriately.

Can we say that man now has enough understanding to lead him to a solution of one of his largest problems, that of war and peace? If these are indeed two separate problems, might it be peace, not war, that is the greater of the two? Does the solution of the problem of war demand first the solution of problems associated with peace? Are the imponderable consequences of peace deterrents to solving the problem of war? Are not the obstacles to agreement frequently resolved only when the consequences of agreement are understood or made certain?

Whatever is regarded as "worthy of change" constitutes a challenge. A challenge is characterized, in part at least, by one's attitude. The prevailing attitude constitutes the environment in which new and pliable human organisms grow and develop. The prevailing attitude, therefore, constitutes a major force in determining further attitudes, not only of individuals but of society, which, in turn, constitutes a still greater force.

Let us illustrate the meaning and significance of this, the nature of the challenge confronting man, and the nature of challenge itself, by reference to a number of different attitudes with respect to relationships between living organisms.

I shall now introduce into this discussion three different kinds of relationships that exist among organisms in nature. One is free-living; another, parasitic; and the third is symbiotic. Free-living implies essential independence and can be destructive for some forms of life; parasitic implies complete dependence and is frequently destructive of the host; symbiotic implies a mutually dependent, cooperative existence without destructiveness. Man is essentially free-living, but he is destructive of the animals he uses for food; he can be infested by parasites such as the tapeworm; and he is dependent upon certain bacteria (in his intestines) which are also dependent upon him.

Although man is a free-living organism, his principal problem arises from the fact that he is also a social organism dependent upon other humans. The destructive attributes of man, which were essential for survival in forests and jungles, have, on his emergence from the jungle, not become wholly vestigial, and at times are exercised in his relationship with other men. Why do some men act with destructive competitiveness while others act with reason, compassion, and cooperation? Should we not expect a greater prevalence of reason than is sometimes observed in an age when man has been so successful in meeting the challenge for survival against the physical elements in nature and against non-man-made diseases?

Will men who are in opposition to one another be able to allow their fellow men to save face so that they may join in a relationship of cooperation or of symbiosis? The term "symbiosis" is applicable in the sense that nations can help each other by preventing the mutual waste involved in preparing for defense or for war. It is inevitable that a reasonably harmonious community of men will one day develop and that war will be an antiquated method for solving problems. Since man needs to be challenged if he is to fulfill the potential that exists within him, must it be war, or destructive competitiveness, that provides the way in which he challenges himself? To provide challenge, must the principal source of opposition be the destructiveness of man? Cannot the unknown and unknowable future provide a greater challenge? Would this not be more conducive to creativity than an atmosphere dominated by defense against destruction? Would not the degree of creativity achieved in symbiotic relationships exceed that attainable where merely a state of coexistence is the goal?

The principles and mechanisms whereby ideas influence attitudes and behavior are beginning to be understood. But before actions are taken the means must be designed to match the ends sought. If conquest, destruction, or annihilation are the ends, then the means devised will be appropriate thereto. On the other hand, if a symbiotic or cooperative relationship is desired, the means will be quite different.

For molecules as well as for men, "chance," "choice," "change," and "challenge," in the forms appropriate for each, have been essential attributes of life. This statement provides a certain measure of understanding. However, it also provides a measure of both discomfort and comfort. We realize the nature of our history from the trends of the past, as it extends into the present and into the future. We see the needs of each succeeding generation for involvement, for commitment, and for challenge, not merely as something good for the body and mind, but as "biologically necessary" for the soundness of man as a whole.

What has been said could be expressed in many different ways. However, this point of view about life, and the affairs of man, is based upon an appreciation of the nature of the processes of life. We are in the early minutes after the dawn of a new era in man's understanding of the universe of living things and of himself. We are beginning to see the emergence of new thought, of new concepts, and of new language that will become incorporated into the fabric of man's existence, and thereby into his protoplasm. These will have come from the advancing knowledge of biology—not biology in the original sense of the word, nor even biology as it is generally understood today. The biology of which we speak is that which will help us understand all that can be comprehended about living things, including man, in the many ways in which each expresses itself. The knowledge and the points of view of biologists in their various aspects will contribute toward such understanding.

What is needed is a change of perspective in man's view of himself and of his responsibility for his "self" and the "social organism" of which he is a part.

Can we ignore the forces within man which are as responsible for his ills as have been the microbes that we seek to control? The models used in considering earlier concepts of man's self-induced ills were created by man, whereas the model of man as part of biological nature, of which man is a recent "model," may serve to explain more, and may provide more useful bases for action, and therefore for life.

By the discovery of "principles" man solves problems that require "understanding" for solution. It is the discovery of the principles upon which the structure and function of living things are based that has produced explosive changes in thought and therefore in man's life in the nineteenth and twentieth centuries. The remainder of the twentieth century will bring testimony to the prediction that the biological view of human life, in all its manifestations, will provide so large a measure of realization as to result in an understanding of ways for reducing man's inhumanity toward his fellow man. Thus the problems of man's "warlike nature," and the problems associated with "peace," may become manageable even if not altogether solvable.

The science of biology and knowledge of biology are as essential for day-to-day living as are reading, writing, and arithmetic. In time it will be necessary for everyone to know and to understand the biological bases for the manifestations of disease and the normal functioning of living systems, including his own.

It is becoming more and more apparent that ideas possess as tangible a characteristic as material substances, certainly as evidenced in the effects they have upon man and his way of life. We live in an era in which printed and spoken words possess a force which bears out the saying that the pen is mightier than the sword. How important it is that the emotional as well as the intellectual basis for thought and action be understood—not that man may control man, but that man may be guided to the fulfillment of his individual potential. This is the challenge that offers the young and those not yet born more exciting lives than those who have lived heretofore.

More can be done about the human condition, collectively and by the individual himself, than ever before. This does not mean that there is not an enormous distance still to go. On the contrary, as the possibilities and potentials appear to increase in magnitude and significance, so do the problems and crises, and it is for this reason that there will never be an end to challenge and opportunity, as long as the young are encouraged and allowed to become men and women who seek the opportunity and accept the challenge it is in their nature to do.

10

The Familiar from an Unfamiliar Viewpoint

Many insights have been conferred upon one science by a related science. This has long been true as human knowledge has steadily evolved. The relatively recent effect of physics upon biology has been as unifying as the earlier effect of chemistry upon biology and the still earlier effect of physics upon astronomy.

Biology differs from other sciences in that it may provide a natural bridge from the physical sciences to the human sciences, which regard man and his humanistic concerns in a way that offers the possibility of uniting the sciences and the humanities, sometimes viewed as distinct and divided, with an ever widening gap between.

It has been suggested by some that to bridge the gap, a humanist, for example, should know the second law of thermodynamics and a scientist should know the works of Blake. Although one cannot question the desirability of broadening the orientation of scientists and humanists, perhaps it would be more immediately useful for a humanist to understand biological systems so that he may understand the biological nature of man, and for scientists—more particularly biologists—to apply their methods of thought to questions of concern to humanists. It would be useful for the humanist to understand the nature of man's evolution all the way from elementary physical particles, and for the biologists and other scientists to view man's scientific and humanistic inclinations, desires, and expressions as part of his biological nature.

What is implied in the foregoing is that man, in order to understand himself, must understand evolution. He must understand the nature of the material of the physical and biological universes. He knows that living matter is composed of elements found in the physical universe. When these combined in particular arrangements, under natural circumstances, they exhibited the characteristics of living things, one of which is self-replication. While environment seems to draw out the characteristics latent within a genetic unit (a replicating molecule), the way in which such a molecule reproduces and constructs the organism itself has begun to be understood. Knowledge of the existence of a molecule that contains a translatable code challenges our desire to understand more about how it works, about what follows therefrom. Such knowledge provokes our desire to know just how this may have come about.

The biology of today—the science concerned with the nature of the structure-function relationship in living things—provides a field of activity not only for trained biologists but for other scientists who are interested in such questions. Moreover, the new biology also provides food for thought for artists, poets, philosophers, and others concerned with human questions.

We have, in a rather roundabout way, arrived at the realization that science needs to become part of the consciousness and conscience of man and that, for the full development of both, science must be incorporated into man's substance, just as the chemical composition of man's blood, which resembles that of sea water, reflects the composition of the environment in which he evolved. So it must be that as man continues to evolve he will incorporate within himself more and more of the environment of scientific knowledge in which he develops and matures.

Biological systems are essentially evolving patterns in which environment, as defined in earlier chapters, may be said to evoke the potential contained within the molecular arrangements in living material. The extent to which "environment," in this sense, is reflected in the substance of living systems is manifest not only in the example of the similarity of chemical composition of blood and sea water, but in the extent to which the substance of living things reflects the effect of a concatenation of biological potential and experiences incurred in the course of evolution. In these ways, the evolutionary potential of the earliest molecular mass of living matter may be said to have been educed by its environment.

From the biology of today we now understand more deeply than ever before the nature of the structure of living substance. We recognize in living systems the existence of a relationship between structure and function in which each structure has a functional purpose. It is also clear that function cannot exist without structure. The replicating molecules upon which life depends depend, in turn, upon their environment to reveal and to develop their potential. This is what happens in the evolution of living things, including ideas.

The science of biology has brought into existence concepts that would not be known if the physical universe alone were studied. Beyond concepts deduced from observing primitive living matter, new concepts are required as the evolutionary scale is mounted and as higher central nervous system activity reaches the level of complexity and functional refinement seen in the mind of man.

The biology of today covers a wide range, from molecules to man. On the other side of this widening frontier is the gain of a perspective and a depth of understanding which promise the development of systems of thought, values, and bases for judgments, as well as a view of man in the universe, which could guide his ethical and moral life to bring him closer to realizing the hopes and aspirations of all of mankind.

The problems confronting man today are far more complex than ever before. This will continue to be increasingly true of the human condition. Until the sun's energy is dissipated and can no longer sustain man, or those forms of life which will evolve and adapt to whatever circumstances then prevail, man's physical survival depends on other, more immediately threatening and limiting factors.

Man is aware that his major problems now are himself as he experiences the self within his own confines, and himself in opposition to the selves of others. These difficulties contribute to the practical problems of existence. If man is to deal with these, he must first understand the nature of his substance and his relationships. The knowledge man needs to allow him to understand the human condition must come from biology. It is inevitable that in the future, if not now, it will be as essential for man to know and understand the laws and relationship of living systems as it is for him to have other basic tools for his existence.

When one thinks in these terms, it is clear that the molecules contained in living material evolved even before the organisms of which they are composed. This evolutionary process applies even to phenomena of special interest to behavioral scientists and social scientists, as well as to phenomena with which poets, philosophers, and other humanists are absorbed.

The question is often asked: "Is psychology a biological science or a social science?" One might ask in reply: "Can one really distinguish psychology, sociology, and biology other than to say that psychology and sociology are in reality subdivisions of the biological sciences?"—just as history may also be thought of as part of philosophy.

Ideas evolve just as do living things. The potential for the development of new ideas exists, and under proper environmental circumstances it is evoked. More often than not this happens when a fresh point of view is introduced—when the familiar is examined from an unfamiliar viewpoint.

The mere thought of an "object" that is concerned with itself in an "objective" manner appears physically impossible. It is obvious that when an "object" becomes introspective, it is immediately changed by this very fact. This is the conundrum which man faces. And yet man cannot escape the reality of this state.

The investigation of the atomic nucleus or of outer space is an expression of human nature. These two dark areas are among the many vast unknowns that challenge the human mind and its need to explore. The navigators of days gone by are now those who explore heights, depths, space, the infinitely small, and the magical wonder of the realm of living things.

It would seem prosaic to refer to the challenge of man's physical ills when a still greater challenge is that of revealing how man can become more compassionately human and less destructive of other humans and even himself. The fulfillment of these hopes may reasonably be derived from a deeper understanding of the science of biology.

II

Man's Biological Potential

Man had a beginning, and we may assume that he will have an end. But the question of his beginning or end is not of as great importance as the question of his present. This does not mean the past or the future is less important. On the contrary, the present has meaning only as part of a continuum. It is from this point of view that we reflect upon man and his "biological potential," a term which conveys the idea of time, change, and evolution. Our interest is more than academic since each, in his own way, is concerned with one or another aspect of this large question.

Man is curious about the mind of man. His curiosity is based not only upon a desire to know for the sake of knowing, but also upon the reason that it will be useful to know. "Curiosity" is a property of the mind of man. Associations are made, concepts are developed, and man eventually sees how knowledge may be used to solve the problems he recognizes. Recognition of a problem comes when one can imagine a situation different from that which exists; man's ingenuity has been applied to influencing nature as he tries to create a world closer to his heart's desire.

Since nature includes man himself, it was inevitable that, as threats to his physical survival became less pressing, man's attention would turn eventually in the direction of his own mind. It was inevitable, too, that the eventual development of a critical level of basic knowledge would bring about the realization of hitherto unrecognized powers. This would permit and encourage realistic thought about problems which might otherwise continue to be avoided.

Questions that are difficult to face seem to defy resolution, and often too easy answers block further inquiry. But this is not the way of science.

Science is man's most recent way of addressing himself to the questions that have occupied him since he developed awareness and the equipment with which to deal with the sensations and reactions evoked by awareness. By virtue of the fact that science exists, its emergence and development were inevitable. The existence of man is also testimony to his inevitability. Nevertheless, we doubt that if a sentient being was present in the universe ten million years ago, it would have been able to forecast the inevitability of man, or of science.

Let us look at our own lives to see when it became apparent that the pattern would unfold as it did. I want to avoid philosophical arguments about predetermination; rather, I want to establish a sense of the limits of predictability as we address ourselves to the question of man's potential from the biological point of view.

In earlier chapters, when I spoke of man's biological potential, I had in mind man's capacity for change—and his own influence upon himself in this regard. I have already indicated that man's view of himself depends upon the premises he chooses and the assumptions he makes. The assumptions and premises I make derive from biology. In view of the earlier discussions, we now see that man incorporates the history of physical and biological evolution. In man a new nodal point in evolution occurred, analogous to the first emergence of life on the planet, and to the first appearance of animal life as distinct from vegetative life. Many other nodal points have occurred in evolutionary time, as, for example, the synthesis from simpler substance of nucleic acids, which make up the coding and decoding information system for heredity, as well as the "invention" of the cell membrane.

The brain of man has now evolved to a point where we are

conscious of the questions which confront us. The brain even possesses mechanisms for dealing, at least in part, with such questions. Concepts, thoughts, and imaginings are the product of that organ. Might not the development of consciousness be analogous to the development of earlier potentialities which existed in simpler forms of life and were contained in structures which evolved into still more complex forms, ultimately to the development of man? If contemplation, abstract thought, science, and technology have appeared for the first time in man, what is the nature and the meaning of these capacities?

Because I believe in the usefulness of biological analogies in thinking about man I have drawn attention to relationships between structure and function in living systems discovered by biologists working at the molecular level. Anthropomorphic descriptions of biological phenomena, rather than being regarded as reflecting man's egocentricity, may be, rather, a recognition that man is an aggregate, at a higher order of complexity, of simpler but still very complex cellular or subcellular systems.

We see the development of science as part of the continuum of the evolution of thought and of methods of thought. The biological "equipment" for this process includes the means for the development of a language expressed in symbols, in writing and in speech. Another, earlier, important, liberating, evolutionary step was the appearance and development of the opposable thumb, of manual dexterity and of technology.

The liberation of man's potential for thought and for technology depended upon the appearance and the evolutionary development of language and of manual dexterity. The potential for thought, which existed in the precursors of man's central nervous system, was further developed, or further educed, by the development of a biological potential for language and for manual dexterity that existed at earlier stages of evolution.

The sharp distinction and separation into different realms of thinkers and doers in the academic and nonacademic world tends to limit interaction among ideas, individuals, and disciplines. Some biologists turn away from questions concerning man that seem as yet too complex for them to encompass, and some "behaviorologists" look askance at the suggestion that biological concepts might provide simple working models for thinking about complex human phenomena. A few "oversimplifications" have already been introduced to provide analogies for such thought. We will now recapitulate and add other examples of basic biological phenomena to further our understanding of man and his biological potential.

We have already described the "enzymatic adaptation phenomenon," in which the synthesis of enzymes is controlled by regulator genes and by which, through feedback effects, the products of an enzyme-activated reaction inhibit further enzyme activity, as would an "enzymostat." More than a thousand enzymes exist in a single cell, simultaneously controlling the interrelated and interacting chemical processes occurring therein. It is because of its general implications that we have called attention to the "enzymatic adaptation phenomenon" as an example from biology that has "human implications." By restating the details of the "enzymatic adaptation phenomenon" I shall present it in a way that is appropriate for the theme of the present chapter.

> Bacterium (B) that does not contain enzyme (E) for substrate (S) will, upon the addition of substrate (S) to a culture of bacterium (B), begin to synthesize enzyme (E). The capacity to make this enzyme and to carry out the particular functions of digesting substrate (S) is genetically determined. Bacterium (B^1), which is also free of enzyme (E), will not synthesize this enzyme even under the conditions favorable for (B). Thus (B) and (B^1) are different with respect to their capacity to make enzyme (E) and to act upon substrate (S). However, this is not the fact we want to illustrate, but rather we wish to emphasize the fact that bacterium (B) does *not* make enzyme (E) *until* it is exposed to substrate (S). Thus the potential of the microbe to produce the specific enzyme is "educed" by the substrate. It is as if the

organism did not reveal its potential for making this enzyme until exposed to the "challenge"—or, stated another way, only under the specific "environmental influence" in which "the cause is the target of the effect." The mechanism is such that it cannot be said that the bacterium has "learned" to digest the substrate, but rather that the capacity of the bacterium to perform in this way was "educed." Those who first discovered this phenomenon described it as "induced" enzyme synthesis. Perhaps we can reconcile the use of the two different words by saying that the substrate "induces" the enzyme synthesis and that the potential of the bacterium to do so is "educed." Neither "induction" nor "eduction" can occur in the absence of genetically determined competence on the part of the bacterium. In passing, I want to mention the interesting fact that the inducing substance operates through the deactivation of an inhibitor which normally keeps the production of the enzyme in check until the inhibitor is suppressed.

It should be evident that when a simple substance such as the appropriate sugar is introduced into a potentially reactive bacterial culture a series of highly complex events is set in motion, and a set of precisely programmed events occurs which reflect the cumulative heritage of this particular organism. The "ritual" of this complex reaction is well established and occurs in a way suggesting the existence of a "cultural heritage" and of "great knowledge" on the part of the bacterial cell, the *anticipatory biological potential* of which is not revealed until "challenged."

I have also referred to another, similar but more complex system, that of antibody formation. In accordance with one theoretical formulation, it is believed that in the same way that the capacity pre-exists for a bacterium to form an enzyme awaiting induction, the capacity to form an antibody pre-exists awaiting contact with the antigen to activate cells "programmed" to form antibodies. Since antibody production sometimes continues long after antigenic stimulation, it appears that the progeny of antigen-induced cells "remember" the initial experience, as if, having "learned its lesson," it continued to "perform the feat" of producing antibody.

I have referred to the effect of a second antigenic stimulus, given at a later time, which produces a reaction of "recall," or, as it is commonly called, a "booster" effect. This exaggerated, seemingly "learned" response is the consequence, in part at least, of an increase in population of cells competent to form antibody, the multiplication of which is stimulated by contact with the "challenging" antigen.

A critical period in the development of the immunologic system is manifest by the induction of either tolerance or intolerance, depending upon the point in time of contact with antigens, i.e., before or after birth. This is evident in the course of the early postnatal development not only of the immunological system but of other systems as well. The example given earlier concerned the limited immunologic capability of animals raised under germ-free conditions. Nonexposure to prerequisite stimulation impairs the capacity of the organism to form elements which, if not available at the proper time in the course of growth and development, give rise to a functionally limited organism. The germ-free animal, upon later exposure to infectious agents, does not have the same capacity for dealing with agents of disease as does a normally raised animal. Thus the development of the immunologic and therefore the full biological potential of the germ-free animal is restricted by the protected or sterile conditions under which it is raised.

A similar phenomenon has been demonstrated in the visual system of cats, in which deprivation of exposure to light, or to forms and shapes, prior to a certain stage of postnatal development results in permanent impairment of function of the visual system. The visual apparatus does not develop fully before birth and unless appropriately stimulated does not develop "normally" after birth.

In respect to the visual system of the cat (and other animals), it is also known that certain cells react to movement in one direction or another and to one shape or another, that some are inhibited and others are stimulated by the same stimulus. Thus, while a single cell reacts to a particular stimulus in a particular way, the organism as a whole reacts to a wide range of stimuli in an amazingly integrated fashion for such a high order of complexity.

By analogy, this suggests that there pre-exists in the different learning cells of the brain a latent capacity for the development and expression of the characteristics and reaction patterns which are later exhibited. In human terms, these genetically determined patterns exist in the brain and are not expressed until impinged upon by circumstances that then develop the skills, thoughts, actions, and total personality which eventually characterize the individual.

It is clear that, for the full expression of the capacity of the organism, the developmental process requires adequate stimulation at appropiate stages of development. Not only does failure of appropriate exposure affect the system involved, but it thereby affects the entire organism. Thus the orchestration of the unfolding of the organism's full capability in the course of development requires the total exercise of all the potentialities for the uses to which they are "destined" in the course of evolution. From this it would appear that by what man does or does not do he influences the characteristics, and therefore the nature, of the predominant type of behavior that exists at a given time and in a given place. If this is the way of the natural evolutionary process, man is now able to influence his pattern of evolution biologically as well as culturally and socially.

This suggests that attitudes and behavioral reactions of many kinds may be constructed in a similar way in the early years of life. The resultant effect would differ depending upon the point in time of development when an incident occurs or an attitude is learned. This would explain rigidity and the difficulty of "unlearning" attitudes and prejudices induced early in life and the need for exposure to realistic and developmentally constructive experiences at appropriate stages in development.

The analogy between the central nervous system and the immunologic system has been applied to the realm of learning. By further analogy with the phenomenon of "adaptive enzyme synthesis" it has been suggested that the mechanisms by which learning takes place may conform to what has been called the "selection" rather than the "instruction" theory of antibody formation. The "instruction" theory implies that all antibody-forming cells are similar in that each cell can be instructed to form antibody to any antigen. The "selection" theory implies that each individual antibody-forming cell possesses the capacity to react only to certain antigens while the animal as a whole will react to a wide range of antigens.

The idea of a "selection theory of learning" suggests that the potential for what is learned is genetically predetermined and is both "induced" and "selected" by capabilities as well as by circumstances. The opposite view is that at the beginning there is a clean slate with unlimited potential for learning in all individuals. If learning is a process of unfolding, then in relation to the question of man's potential it suggests the possibility of educing what might be thought of as the "more desirable" human attributes and capabilities, and of not encouraging the "less desirable."

The foregoing may be summarized by saying that as man goes through the plastic period of his development, during the prolonged period after birth, his mind and emotions are shaped by the experiences and ideas to which he is exposed. The processes involved may be likened to the examples cited. Experiences will similarly influence character, values, and choices and, in turn, determine the causes and goals to which he becomes committed. It is as if the limits of possibilities were genetically determined and existed in protoplasm anticipating demands that will arise.

We are beginning to know enough of the workings of biological systems to be able to deduce general principles which might be applied toward our understanding of man and, particularly, of those attributes and activities of man in realms that are not usually approached from the biological point of view.

Science, it is said, will not be able to explain everything. This is not the point. The possibility that is proposed is that the comprehension of living systems might help us understand more about man creatively, and artistically, as well as physically.

If man is looked upon simply as a machine, he is a remarkable machine, able to produce ideas and able to respond to environmental influences that bring out possibilities which we had no other way of knowing about. This is analogous to the discovery of the existence within a microorganism of a latent potential for enzyme function which is revealed only when the organism is exposed to a particular "environmental influence," or that of antibody formation in an organism exposed to an antigen which it had not previously encountered. These effects involve the evoking of a response through the suppression of repressors that are part of the mechanism of homeostasis—part of a mechanism for conservation of energy and resources for use primarily for functions called upon in the present, and for the activation of those required for a state of readiness in the future.

12

Education—For What?

Man has come through a long and complicated evolutionary past. While evidences of his subhuman heritage still persist, he reaches to explore a human future that is essentially uncharted.

Man differs from other animals in the long postnatal period during which physical, emotional, social, and intellectual maturation can occur. It is during this period that metamorphosis proceeds from animal to human. The degree to which this occurs successfully is a measure, in part, of the nature and influence of the nurture and the education to which the individual is exposed in the early years of his life. Nurture of the young, and education in later years, are largely man-determined, and are *widely different* for *different cultures*.

The biological potential of man at birth is far greater in relation to his mature state than it is in species with a relatively short postnatal developmental period; in the latter the potential is essentially fixed at birth, or soon thereafter. Such species undergo evolution very slowly; mutations or genetic variations, under the influence of environment, become additive with time; and differences become marked only over very long periods. When we consider human evolution broadly, including nongenetic, *culture-induced* evolution, man, in contrast to subhuman animals, has had a marked and rapid influence upon his own evolution. This has been effected through his capacity to influence his environment, and through his *cultural heritage*. Through a variety of pressures that he has applied to himself, capacities have emerged, the potential for which was latent within him.

Since the inborn pattern of each individual allows modification after birth within limits set by his genetic endowment, it should be possible to treat the early modifiable stage of man in a way that may increase the probability of bringing into greater evidence what might be considered man's positive human qualities and capabilities.

The point has been made in earlier chapters that sensory impressions, experiences, and ideas produce effects through the impressionable organ—the brain—during the period before maturation is completed. In this way, attitudes of tolerance, or prejudices, are established, just as the pattern or prior immunologic experience affects later immunologic "attitudes," whether they be protective, allergic, tolerant, or intolerant. Thus more opportunity to exercise constructive attributes, and less opportunity to exercise destructive attributes, would be expected to influence an individual's predominant attitude and orientation. Hence we see the importance of timing, as well as of the nature, of early educational experiences, for the establishment of knowledge, attitudes, values, and goals.

We have referred to a "selection theory of learning," meaning that the potential for what is learned is predetermined, rather than that at the beginning there is a clean slate with unlimited potential in all individuals. The recognition that learning is a process of unfolding bears on the question of man's potential and on the possibility of educing the "more desirable" human attributes and capabilities. Whether this can be accomplished by design and plan is difficult to answer (although it would seem to be possible theoretically), but the reverse has occurred through default, or ignorance, or through absence of discipline, with the emergence of "less desirable" characteristics.

We need not be reminded of the empirical fact that habit formation is most easily induced early in life, and that the nurture and education of the young are important. It is now known that some types of learning can be introduced earlier than had been assumed and that concepts, abstractions, and mathematics, as well as second languages, can be learned as early as three to seven years of age.

The foregoing may be summarized by saying that man goes through a plastic stage of development, during a prolonged period after birth, during which his mind and emotions are shaped by the experiences and ideas to which he is exposed. This influences not only his knowledge, but his character, his values, and his choices and, in turn, determines the causes and goals to which he becomes committed. The limits of possibilities within him which are genetically determined exist in his protoplasm, anticipating demands that will arise. Education and experience bring out the pre-existing possibilities that would otherwise remain latent, or would never be expressed, unless opportunity occurred to develop the potential that emerges.

We are at a critical stage where we must ask, as did the ancient Greeks, "What are we to educate for?" As the values of the Greek city-state changed from personal qualities of nobility of character to those of intellectual attainments measured by success, not always dependent upon impeccable character, the goals of education changed.

It is strange to contemplate the vast changes that have occurred since then. Modern man finds it possible to master the intricacies of putting men into space and on the moon, and yet he seems unable to solve the problems of poverty, starvation, and war, nor even the seemingly minor, but no less significant, wars within communities, between neighbors, who seem to wish neither to destroy each other nor to accept each other. Nor has he been able to resolve the wars within himself.

It is more important to find the right questions than to provide a priori answers if we are to seek to establish a quality of character that could justly be referred to as "noble" as the goal toward which education should be striving.

What is meant by noble? The dictionary speaks of "exalted

moral character or excellence." Can we offer in place of this, as a workable meaning of nobility, "doing what is right for the right reason"?

The potential for nobility, as well as for brutality, must be biologically endowed and therefore is present at birth. There must be a time in life when man's nobility could be evoked and then strengthened, a time perhaps still under the aegis of the parents and educators, who themselves must be noble and who later may have to battle the influence of persisting ignoble elements. Where, therefore, are we to begin and when, in terms of the life span of each successive generation?

If school is a part of life itself, and not merely preparation for life, then should we not expect nobility there to serve as an example? Can nobility be taught, just as reading, writing, arithmetic, and biology are taught?

Perhaps we should seek a preventive against the human disease which we might call "brutality." What could we "inject" to induce a positive effect, how can it be done, and when? If it is done as early in life as possible, it will still be too late for those already crippled. Many have not yet been so maimed, however, and all the young are susceptible. Since nobility rather than brutality is what is desired, we need to determine how it may be induced.

We have a long way to go. Let us not deceive ourselves, but recognize the difficulties of the struggle.

Although the ideal may never be attained, we must, nevertheless, establish a goal based on a realistic appraisal of the nature of the problem. Man is a biological organism not yet completely evolved, or not with sufficient uniformity for his brutality to have become vestigial enough that his constructive biological potential can be easily and fully brought out. This is a process which will require the passage of many more generations before we know whether or not the ideal can be attained.

An attempt is being made in this book both to reveal the nature of the biological processes through which such effects may be mediated and to suggest a theoretical basis for evaluating courses of action and trends as we seek basic rather than expedient educational solutions to man's most pressing problems. The educational problem of the present—in all cultures—requires that we teach the young the greater importance of acquiring a constructive method of thinking which keeps the mind open to the discovery of new aspects of reality in a dynamic approximation of truth than of equipping them with ideas that resist change. It is also crucial to equip the mind to think about increased cooperation rather than increased competition. We must be open to such changes, learning about the nature of science and culture, and about the value of both for man's survival as well as for his personal satisfaction and fulfillment.

13

Violence and Counterviolence

Struggle and expressions of violence on campuses and in communities in the United States are not new but have recently increased. As a result, a great deal of energy for growth has been locked in contests from which release is now not simple.

When hostilities are strong and nearly paralyzing, we sometimes look inside ourselves for a way to put events of the day into perspective and to provide relief from anxiety.

As for the epidemic of distress, destruction, and violence, can we by using analogy to the process of disease in man be helped to understand the outbreak of restiveness among youth?

The existence in society of some kind of malfunction or "disease" is evident not only in unrest among students but in other manifestations of disorder—many of them violent. Violence suggests that irreconcilable differences have reached a breaking point. When violence begins, it often grows and extends until some of the cause is destroyed. In the course of violence part of man is destroyed. The growth and persistence of violence signify a serious basic disorder, the cause or causes of which need to be known lest we merely suppress the signs and symptoms and fail to eliminate the cause, which might then be revealed in some other way that was even more serious and insidious.

A physician confronted by a similar problem in a patient begins by trying to distinguish between effect and cause, between secondary and primary events. He tries to distinguish between effects directly attributable to the causative agents and those attributable to the activity of the defense mechanisms of the patient. He directs his efforts toward suppressing the causative factors without hindering the mechanism of defense unless the latter, by overreaction, becomes harmful to the patient.

A fever, for example, is a healthy sign that the body is combating a disease. A physician usually does not suppress fever unless it becomes so high as to be harmful to the patient. He concentrates, instead, on the cause of the fever.

As applied to disorders in human relationships, each side in a conflict regards the other as criminal and feels himself the victim. How, then, can distinction be made between cause and effect, or between criminal and victim?

Since violence evokes counterviolence, it can be regarded both as an effect and as a cause. While it is generally desirable that violence should end, man has not yet discovered how to control the switch—more easily turned on than off.

War is an institutionalized form of violence. Students, today, in general demand an end to it. The continuation of institutionalized violence, such as the war in Indochina, has been in part the cause of some of the violence on campuses. There are other reasons which must be explored to determine to what extent campus violence is due to the special sensitivity of youth and of others who recognize that the rules by which we live are not always in consonance with the laws of nature, who believe that the rules of man must be brought into accord with the laws of nature. So long as double standards exist, such as the double standard about violence, antiviolence may become counterviolence, and thus violence and antiviolence eventually become indistinguishable parts of a newly created situation.

Except for war, essentially all forms of violence are "illegal" and are punishable in one way or another. War persists as a noncontrollable institution which man tries to regulate. But if "nature," residing in the freshness of youth, is beginning to take a position against this "officially accepted" form of violence, then the protest of youth against war is a form of health rather than a manifestation of disorder or disease. This is not to be construed as implying that the violence of youth is to be condoned, or that counterviolence is justifiable by those whose role it is to protect society. The chain-reaction effect of violence and counterviolence is evident in the sequence of events that in several instances has led to the deaths of university students.

As events in a chain are further and further removed from the primary cause, it is increasingly difficult to distinguish cause from effect, as in the case of a physician who treats a disorder with many secondary consequences and who finds it difficult to do more than treat symptoms as they appear, just to keep the patient alive and as comfortable as possible.

The society of man, while suffering from ill health, will survive for a long time to come. Despite those who predict early doom, it is likely that the human species will persist long into the future. Nevertheless, the question is not merely survival but what life will be like and how healthy it will be.

If man is able to bring under control elements *in nature* that threaten human life, he should be able to bring under sufficient control their counterparts *in the human realm.* Since individuals, as well as cancer cells, can be lethal to man, much knowledge will have to be acquired and much work will have to be done to bring about a measure of control over growth both of population and of greed in man, excesses of each of which might be thought of as cancers *of* man, analogous to the problem of cancers *in* man.

Man needs to know more about the order of nature as expressed in himself so that he may revise the rules by which he lives, individually and collectively, to design the kind of life he desires, and to act responsibly to make this possible.

We need to unify the positive elements and philosophies in society so as to create a strong, positive movement that will attract the like-minded. Man must unlearn a great deal in order that he can start to learn anew the values and practices that must prevail for survival and health.

14

Disease and Counterdisease

Drugs are taken by some for the purpose of producing transient feelings of well-being. Such drugs are also taken for relief from the pain of the vicissitudes of life. Drugs of this kind become part of a way of life, more powerful than the will of the individual to overcome. If addicted, such persons become victims of a disease, the manifestations of which vary from mild to severe, from acute to chronic, and from curable to fatal.

Let us think of drugs that produce these effects as etiologic agents of "drug-taking" diseases and as analogous to bacterial or viral causes of disease. Different types of viruses are more or less virulent and give rise to different clinical manifestations; similarly, various types of drugs are more or less harmful and cause a variety of syndromes. Just as certain viruses and bacteria are transmitted by vectors, or carriers, so drugs may be thought of as transmitted by "carriers," who themselves are often "infected" in the sense of their using, or being addicted to, drugs.

If we use the analogy further, then the reaction of society to "takers" or "carriers" of drugs might be thought of as analogous to the immunologic reaction to an agent of disease.

The immunologic system normally reacts "protectively" to the recognition of a foreign invader. For example, on or about the tenth day after exposure to measles virus, it does so with the development of the rash and fever that accompany the immunologic reaction to the measles virus.

There are other instances in which an exaggerated immunologic, or self-protective, reaction results not merely in an acute response, as in measles, but in severe and chronic disease, as in rheumatic fever, rheumatoid arthritis, and chronic nephritis which follow certain streptococcal infections. This occurs as a result of the normal protective reaction to eliminate streptococcal products. This causes destruction of the normal heart, joint, or kidney tissue to which streptococcal elements are attached or which are simply immunologically similar. The effect is the same as that of an "autoallergic" reaction.

Society seems to react in an analogous way when a "drug-taker" or a "drug-carrier" is identified and apprehended. In reacting morally with a kind of "rash" and "fever," and legally with a kind of "autoallergic" response, the reaction that is ostensibly "protective" paradoxically damages the lives of young, innocent victims and minor offenders with ineradicable criminal records. Thus many victims of the "drug-taking epidemic" may be more severely and permanently damaged by the "protective" reaction of society than by the drug itself, provided the drug in question is not of the permanently addicting variety.

This is the effect if drug-taking is treated not as a disease but as a crime. Other mental and behavioral disorders are also still viewed as "crimes," which might better yield to control measures if treated as diseases.

This view of the drug-use problem is presented to suggest an approach to control. In the absence of effective measures for eliminating the "etiologic agents" (drugs), or for distinguishing "victims" from "carriers," or for effectively controlling "carriers" and those who create and profit from "carriers," it is suggested that a form of "immunizing" education be developed, aimed at the as yet "uninfected" susceptibles. If such "immunization" is widely enough practiced and if a significant proportion of the population is so "protected," the disease would fail to spread for lack of sufficient susceptibles. Immunologically and epidemiologically this is known as "the herd-effect," by which even the unvaccinated portion of a population is effectively vaccinated against an infectious disease. It is also suggested that laws should distinguish between victim and criminal. The secondary problems created by drug-taking, through criminalizing victims, is analogous to a kind of "auto-allergic" reaction (against victims) which creates new problems rather than solving existing ones.

The precise nature of the measures needed for an "immunizing" education would have to be developed by specialists both in the art of persuasion and in the art of education. It may not be sufficient to discourage by fear of harm; it may be necessary to provide the expectation of greater satisfaction in *not* taking drugs, and greater pleasure in solving life's problems *in other ways.*

Drugs have long been available. A new pattern, however, has developed. It is not enough to say that the cause is attributable only to those who traffic in drugs for financial gain; there have always been exploiters who feed upon human weakness. The possibility must be considered that taking drugs may be a sign of a more basic malady. It is conceivable that drug-taking may be related to such excesses as overcrowding of time, of space, of expectations, etc. From the viewpoint of the survival of the species, though not of the drug-takers themselves, the use of drugs may have a paradoxically positive effect by reducing the number, or effectiveness, of those weak in self-control and self-discipline who might otherwise be engaged in other forms of excessive behavior. The "wisdom" of nature may be operating through the epidemic occurrence of drug use, to reduce excessive productivity in excessively productive societies.

Thus the drug-use epidemic may be acting as a part of a negative feedback phenomenon the effect of which is to reverse the tendency toward further accumulation of the products of "desire" and "purpose" which have resulted in excesses from which society now suffers. The effect would be of value to the species and to the as yet unaffected individuals in reducing the crowding of space, of time, and of the psyche of man. The drug-use epidemic is sufficiently recent in origin as not yet to be seen in its true meaning. The reduction of drug use, if it were to become possible, may simply expose a still more basic underlying human disorder of which it may be but a symptom.

We have tried to suggest that the epidemic of drug-taking may be an expression of the reaction of individuals en masse to the circumstances of life in a way that reflects the operation of species-preserving forces in man even at the expense of some of its most valuable individuals. Thus we must look more deeply into this and other problems to determine whether or not species forces as well as social and individual forces are at work, so that we may then be able to develop appropriate remedies, whether for the diseases of the individual or for those of the species, rather than a nonremedy that creates a new problem without resolving the old.

Whether drug-taking is an effect of the personal or social circumstances of an individual life or a ruthless correcting of nature's going too far in the evolution of man's mind, it will be necessary to try to understand and deal with the problem with the same kind of penetrating and disciplined thought scientists have used in probing the nature of the processes of disease which have been brought under control.

Greed may be one of the diseases in man that needs to be controlled. This would require a major reconstruction of man's view of himself individually as part of society and a new view of the organization of the human species. For the survival of the species and for the fulfillment of individual lives free of pathological greed, new values and new ethics are required. The essential requirement for such a value system would seem to be the creation of an *identity* of interest, rather than a *conflict* of interest, between the individual and the species, between the citizen and society, and between nations and mankind. This basic issue must be faced lest men in their greedy competition for a way of life destroy themselves and their planet.

Wherein, then, lies hope for man? It would seem that hope lies in all of us, individually and collectively—through the augmentation and strengthening of the primary center in each human individual for self-expression as well as for self-control, and through humanizing the control measures required by society to deal with failure in the self-controlling mechanism of the individual. The individual has a biologically rooted need to be responsible not only for his own needs and deeds but for those of society, and of his species. The widespread disorder of individuals, described as "a search for identity," and revealed in the battle against "alienation," is a manifestation of a need of the individual for relating both to his *self* and to his *species*. This complex and difficult problem requires the most careful thought and consideration for developing appropriate experiences in education and training in the "course" of living.

If life is a process of discovery through unfolding, then a basic imbalance in some minds may be the result of our expectations that the course of life should be known, fixed, and secure. The opposite seems to be the natural state, in which life unfolds in mysterious ways and is discovered in the course of living in uncertainty. If in human life joy lies in self-discovery (and I think it does) , the self will never be discovered and life will be without joy unless it can be allowed to unfold with all the uncertainties and therefore anxieties this involves. If we try to obliterate the anxiety attendant on not knowing how life will unfold, and dull the sensations required to reveal the self, then growth soon ceases and the self soon dies. Healthy anxiety is part of the nature of life itself. It functions to signal excesses and imbalances which we are called upon to respond to. We must learn to distinguish healthy from what might be called unhealthy anxieties.

I have been trying to say that it is necessary now not only to "know thyself," but also to "know thy species" and to understand the "wisdom" of nature, and especially living nature, if we are to understand and help man develop his own wisdom in a way that will lead to life of such quality as to make living a desirable and fulfilling experience, not one to escape from by the abuse of drugs which, in themselves, can have value when used appropriately for specific and relevant purposes.

15

Health as Wholeness

To develop the health of man it is necessary to understand what it means and how it is developed. Once understood, the next step is its improvement. But before we can devise ways and means, we must first understand what must be accomplished.

To achieve a state of health, in the broadest sense, it is necessary to understand not only the development of cells and organs but the development of the individual and the species as well. And if the health of man is to become fully manifest, it must prevail not only in the individual but in mankind as well.

Health is wholeness, and sickness implies impairment of parts of the whole. Distinctions must be made and the relationship understood between the parts and the whole, so that attention may then be directed to maintaining or to repairing the health of each appropriately.

Attention must be drawn to the distinction between the parts and the whole as well as to the relationship between them in order to reveal the meaning of health as wholeness. While so much thought is being devoted to the environment and to ecology, it might be useful to put into perspective the health of man himself as the essence of our concern. The problem of man's health will not be dealt with by solving problems of the environment or ecology. Even if these are alleviated, it will still be necessary to deal with the health of the person and the health of the species.

In this sense, not only the human individual but the human species needs to be studied from the viewpoint of health or wholeness. A science of health, as distinct from the science of medicine, is needed to deal with the problems of human development responsible for a large part of the misery and despair of mankind, manifest in man's psychosphere and sociosphere, as well as his biosphere.

A great deal of the pathology of later life develops as a result of unhealthy experiences in the early years of life about which more must be known. More must be done to prevent the waste and malformation not only of children but of the adults into which they grow and develop. The pathology of man, in body or mind, can grow so large as to become an excessive burden.

As we think of the human problem in this way, we see it as a problem of the whole and its parts—the whole being mankind, the species, and the parts being the individual persons. We can then comprehend the general as well as the particular in the need for achieving health or wholeness, and develop a concept of the way in which the individual and mankind fit together and the nature of the whole of which they are a part. This means that we must first develop our sense of the patterns of order in human relationships just as these patterns have been revealed in the physical and biological realms.

Our awareness of biological patterns marks the beginning of knowing what needs to be done for one's own benefit and for fulfilling a collective purpose. This is a most difficult undertaking. Yet a beginning must somehow be made. This implies a degree of evolution of consciousness such as does not yet exist, and which first comes about by chance as the forces of life play upon individuals, who then consciously develop such understanding. Such consciousness will help in guiding and teaching others to facilitate their own development and evolution. Only then will it be known to what extent it will be possible to improve the health and quality of human life as measured by the fulfillment and satisfaction of the capabilities of each individual in terms of what he can do best himself and collectively as part of mankind. To realize this, it will be necessary to establish patterns in the young child that will help him develop to his fullest, in the way most satisfying to himself and in relation to others.

The definition of role in life is, perhaps, one of man's most important necessities. It will become still more important as the speed of change diminishes in the future and the multiplicity of opportunity increases by virtue of the multiplicity of the needs of mankind. Each individual will have to choose his set of roles in the scheme of things from birth to maturity and from maturity onward, and with many changes in direction and emphasis. We will need to repattern our lives if we are to see the fullest and most effective development of constructive individuals and reduce the accumulation of destructive, unfulfilled, and dissatisfied individuals.

Man differs from other animals in that his behavior is to a large extent beyond genetic control. It is for this reason that consciousness is required to establish the equivalent of the unconsciously operating control and regulatory mechanisms of nature, such as govern the highly ordered complexity in healthy systems of molecules, cells, and organisms. Cultural man, free of genetic regulation, at times behaves as a kind of uncontrolled cancer. At other times he manifests behavior, including violence and competition, such as one sees among unrelated species which live at the expense of one another, or compete with one another for the same food or territory. When man is seen in this way, the paradoxes of human life and the basis of the conflicts among men become more comprehensible. It suggests that if man is to solve this problem by coalescence, he has to move consciously to form an organism of mankind as part of an ecosystem related to a purpose.

It is necessary to think about man in terms of a highly ordered, differentiated system of individuals with widely divergent temperaments, talents, tastes, and interests. If we think in these terms of the relationship of individuals to each other and to the whole, we can comprehend the importance of health in the human realm not as a question of one part to the exclusion of another, or of one part against another, but as the health of the whole. It should be evident, from this point of view, that the kind of exclusiveness and competitiveness that have prevailed, which tend to exhaust rather than conserve resources, are a manifestation of an excess that is counterproductive and that will, in time and of necessity, have to be abated. It is likely that men will find it necessary to move cooperatively toward each other, and do so in conscious ways before they are faced with the predicament of trying to repair damage that may be irreparable.

If the process of mankind's development is looked upon as analogous to the unfolding process in the lives of individuals, and to the developmental process in living organisms, then by consciously thinking about man in these ways we have a basis on which to anticipate the problems that might arise or that might be avoided. We may then acquire the knowledge and understanding that would influence, now and in the future, the development of a state of health such as is imagined to be possible. This does not mean that man will be free of disease or spared death. On the contrary, disease and death are an integral part of health and life. They are the boundaries that define limits and provide guides for meaning and understanding.

The changing feelings of satisfaction and dissatisfaction in the course of individuals' lives must also be understood as a guide to seeking the real satisfactions for which the unreal are sometimes mistaken. While a life of perfection can never be ordered, the boundary conditions within which one can experience the fullness of life's satisfactions can be understood. But for this to be possible for more than just a few requires a deeper *knowledge of the requirements for health* in the sense not only of *freedom from disease* but of the *wholeness of man in all the dimensions of his being.*

16

The Mind of Man

Man is his mind. His interest lies in the functional content and possibility of his brain. When we stop to think about it, the newest, most important and interesting step in evolution occurred with the development of the cerebral cortex, when the mind of man began to function as we know it. As for the body-mind dualism, the body and the mind are two parts of a single whole just as the germ plasm and somatoplasm are two parts of a single entity. When we speak of human behavior and the functioning of man as a tool or as a weapon, we are really speaking of the workings of man's mind. There are healthy minds and sick minds. There are times when each of us is likely to think his mind is healthy while others are sick.

An interest in the functioning of the mind can be justified for its own sake. However, the mind is also the instrument for examining itself. If we were to comprehend the mind in terms of differences in functional patterns, we might then be able to improve its healthy functioning and its further development and evolution.

Increased consciousness is required for man to understand his own mind. Many today are trying to improve their "self-awareness," and "self-perception," through "sensitivity training," "body awareness," and even through the use of drugs. There is increasing awareness of the distinction between and the relationship of the subjective and the objective, the non-rational and the rational, the unconscious and the conscious. In short, many people are becoming self-conscious, or conscious of the self, and seeking ways and means for better understanding of self and others in order to function more effectively in relation to both.

What is needed is a way for the mind to perceive itself and the minds of others—a way of looking at one's self as well as others. This is becoming increasingly important for dealing with today's problems. Earlier in evolution man's mind served effectively in dealing with problems of physical survival. At this point, however, the mind is not needed or used in the same way as in earlier evolutionary time. It now requires other uses to remain healthfully and constructively engaged. If not, it can become destructive because its need to be occupied is so overriding.

The mind appears to function as if it were an independent entity housed in and using the body to serve "its" purposes, constructively or destructively. If it were possible to classify minds and determine the reasons for the dominant pattern of behavior, we might then learn how to influence the use of the mind in constructive ways.

In whatever way the mind is to serve in the evolutionary scheme of things, its emergence was of major significance in evolution, representing a nodal point not unlike that which marked the transition from nonliving to living forms, from vegetable to animal, from aquatic to terrestrial, and from those earthbound to those airborne. Man's mind is the "cause" of what we identify as human, including the capacity to consider the mind itself as well as the universe outside itself.

Many scholars have made important contributions to our knowledge of the functioning of man's mind. Men have long been trying to work with it and influence it. Now the mind of man is turning upon itself; its own examination has become a matter of interest, not merely to the few whose profession it is, but to each who is aware of the mind he possesses. Without thorough understanding, each tries in empirical ways to deal with subjectively manifest disturbances, some of which arise when the mind is not compatibly engaged within the evolutionary scheme.

The frequency of mental pathology seems to increase as the engagement of the mind is diminished in relation to the needs for survival. Therefore, each individual must understand enough about the mind to bring about its healthy development and its fullest expression. This, in turn, would afford the fullest measure of satisfaction and of effectiveness in relation to self and to others. In a sense, one of the most important and powerful parts of man's body, partly under voluntary control, i.e., his mind, is not well enough understood to be used consistently as a tool for improving his life. Rather, it is often used unwittingly as a weapon, in the attempt to control others or to keep from being controlled.

When there were masters and slaves, the physical bodies of men were used by other men. The same has been true in respect to men's minds. The master-slave relationship implies a need for, or an imposition of, direction. It is becoming increasingly apparent that an important source of direction must, for each, come from within himself. In the past the more important source of direction was thought to come from without, from some supernatural force; now it appears that the source lies within the mind itself. How many of us possess minds that can become sufficiently conscious of self and yet sufficiently disciplined to serve the species as well as the self in the ways here implied?

The mind does not reveal itself to itself except indirectly. The mind must surmise what the mind is, what its elements are, and how it works. Then it must test such ideas until it solves the puzzle of its own existence, its own nature, and its own functioning. The mind must interest itself not simply in classifying the manifestations of its behavior; it must become interested in its own anatomy and physiology, its own developmental biology and evolutionary significance. The individual mind needs to become conscious enough of its own nature, of its own workings, to be able to increase conscious control that will permit the making of judgments based upon values derived from nature which can be tested both for their objective and subjective validity and in evolutionary terms.

The purpose of this point of view is to emphasize the need to see the problem of man in a way that might be used to prevent the further development of imbalance and eccentricities in man. If we think in terms of analogy to other species whose bodies became so overbalanced in relation to the mind that they met the fate of the dinosaurs, man's mind also seems to have developed dinosaurian qualities. Man's mind seems to have evolved to a point which threatens to overpower his body. If his highly developed mind is not put to constructive use, an enormous amount of human waste will ensue, with all the tragic pathology attendant thereto. Since man's capacity for sentiment is also highly developed, he can become preoccupied with taking care of the increasing amount of human waste resulting from his failure to understand and use his mind properly and effectively and his misuse or abuse of it in ways conducive to decay and disintegration rather than to further evolution.

The foregoing recalls the problem of poliomyelitis, which arose in the wake of improvements in hygienic conditions as part of the advancement of civilization. Now another kind of crippling seems to be associated with advancing civilization. In this instance the problem is of a different nature and far more difficult to arrest. Nevertheless, it must be identified for what it is if means are to be found to cope with it preventively, just as polio had to be dealt with preventively and not curatively. The new form of crippling which involves the minds of young people must be prevented lest they remain handicapped for life as were the victims of virus infection.

A great deal of knowledge was required before it was possible to deal with polio. Much had to be learned about viruses, about cells, and about the immunologic system, as well as about the pathologenesis and epidemiology of the disease. But even before such knowledge could be used for prevention, it was necessary to learn how to save lives and reduce crippling during the acute phase of the disease so that many could recover to full functioning; of those who remained crippled, some managed to contribute significantly to their own lives and to the lives of others.

Analogy suggests that it will be necessary to understand how to deal with the acute phase of disorders of minds facing uselessness—highly developed minds suddenly without use or purpose. This is the nature of the affliction of some of the brightest and most socially and economically "advantaged" youths. The reasons may not be too different from those prevailing when some of the most vigorous, the most socially and economically advantaged fell victim to polio. They had advantages that spared exposure to disease early in life and remained unimmunized, susceptible and vulnerable later in life. The new disorder is likely to extend as civilization advances in the same way as epidemic polio followed in the wake of advancing civilization. Many parallels are suggested between the current afflictions of the youth and the afflictions to which we addressed ourselves a generation ago.

Will man become a package of pathology or remain a package of potential? It would seem more worthwhile to learn to devote one's life to the development of potential rather than to nursing pathology. It will be necessary to analyze further what may be done to prepare each individual to know enough about his mind and his body as instruments for increasing satisfaction in life rather than for housing misery and crippling. This is the reason for trying to understand the nature of the mind so that the individual mind may then comprehend itself. Knowledge of the nature of living systems may contribute to an understanding of the nature of the mind, which is also a living system, and which must have an evolutionary role as well as a specific function, such as, for example, the immunologic system.

What may be needed now is an empirical discovery of the way the mind may be developed and constructively used, similar to the discovery that enlightened Jenner when he realized how to control smallpox even before any knowledge existed of viruses or of the immunologic system. We may be in a similar stage of development in relation to the "epidemic" of disorders of the mind now manifest in many unmotivated young people who are developing an anomie, while others are searching for ways to fulfill themselves. Many allow their minds to become completely preoccupied with their subjective feelings. What may be needed is an activity or a purpose with which to be engaged while pursuing a search for a general understanding of human life, including the subjective part of life.

Thus there is a twofold problem: finding ways and means of engaging the mind constructively and finding ways of thinking about the mind so that its workings can be known sufficiently to engage it in the evolutionary scheme. Some are interested only in using the mind for practical purposes, not at all in understanding it. Others are concerned only with trying to understand it or manipulate it for subjective purposes, and show no interest in using its intellectual power.

These seemingly alternative positions suggest a need for the development and use both of the intuitive and the intellectual capacities of the mind in a unified and balanced way.

It is for this reason that I speak of purpose at the same time as I speak of increasing awareness or consciousness. If we are to influence our own individual destiny and the destiny of man, we must combine, rather than choose between, alternatives if doing so possesses an evolutionary advantage.

17

Man Unfolding

Man is the possessor of particular sensitivities each of which feels and signals too little or too much. These compose the fine structure of the total sensibility of the whole individual. It is necessary that this system be in good order for the individual to function effectively and satisfyingly to himself and to others. It is easy to impair this mechanism but difficult to repair it. It is as necessary to be trained to sense it and use it as it is to be trained to use such capabilities as playing the piano, composing music, playing tennis, mountain climbing, or anything else, including relating with one's inner self and with others.

The greatest amount of pain and discomfort psychically and physically arises from the disharmony felt when individuals come into conflict with each other or when an individual is in conflict with himself. A host of built-in senses are disturbed, senses which are part of an ordering system in man that may be analogous to the proprioceptor mechanism, or position sense. Just as our sense of position or balance is exquisitely responsive to excessive abundance or deprivation, so the relationship-mechanism to self or to others, or to what one is doing, is remarkably sensitive to excessive abundance or deprivation.

There are many different kinds of sensitivities in time as well as in space, in relationship to self and to others, which when impaired or disordered are responsible for difficulties with one's self and with others. No one individual is, at all times, perfectly attuned to his own self and to others. For this reason difficulties exist more with some as compared to others, just as some function better in one area or another while others function well over a broad front and still others function poorly.

It would be interesting to know enough about these human patterns and to know at what points in early life they are distorted or impaired in a way that deprives the individual of the opportunity to develop to his full capacity, appropriately for living harmoniously with himself and with others. While not all individuals are equally endowed, subsequent impairment, or "crippling," can occur because of something that happened, or did not happen, in the course of individual human development. If we want fewer impaired humans, it will be necessary to study the fine structure of the developmental sequence of the attributes and characteristics that are highly developed in man so as to enhance health and reduce pathology in these realms.

The "human potential" movements are made up of people who want to improve themselves individually and in relation to others. They are different from those engaged in social reform. The difficulties of which each is conscious lie in both realms, and, as yet, we have an incomplete understanding of what is required for improving either man's relationship to man or man's relationship to himself.

We may have as much to learn from our imaginations as from past experience. It is easy to assume that our difficulties are due entirely to others, or to experiences in early childhood. Nevertheless we need to develop ways and means of improving relationships if we are to establish balance, equilibrium, and harmony in relationship to self and to others so we can function with greater spontaneity, greater harmony, less fear, and therefore less anxiety than now prevail.

Some people seem to behave as if they were more secure than others. Some are more sensitive than others. If pathology is to be reduced and health enhanced for the individual and for mankind as a whole, it must be seen from the viewpoint of all individuals in all activities, not merely from that of one particular group or another. It seems to be essential that we understand the requirements for developing ways and means for protecting the sensitive and vulnerable human mechanisms so they may develop and function optimally for whatever evolutionary purposes they are to serve. This is essential if man is to develop his humanity fully, as he goes through the maturing sequence in life and as his relationships change in the unfolding of the being which came into existence at the moment of fertilization and comes into flower through the many seasons of life.

We do not usually think about man with the same degree of need for understanding his fine structure of development as we do in attempting to understand cellular differentiation, using the concepts of modern biology. Our success in understanding the nature of the relationships that exist in the control and regulation of living systems strongly suggests what is needed if we are to develop sufficient and appropriate understanding for improving the human condition. It is easy to be hortatory, but for man to develop harmony with himself and with others and to prevent, or to deal with, the conflicts within himself as well as between nations, it is necessary that man be seen from an evolutionary point of view. We must apply biological ways of thought and study.

Biological knowledge and understanding can be drawn upon increasingly for dealing with the problems of man, perhaps in ways that will be more powerful and more effective than those heretofore employed. Efforts at solutions of human problems have been stated in terms of subjective belief when, in fact, the questions are far more complex, and cannot possibly be resolved by what is believed to be *the* answer at any given moment in history. The biological way of thought is not an answer but is a way of finding answers. It is a way to examine and order questions so as to be able to deal with them appropriately.

If man is to contribute to his own development and evolution, it is necessary for him to know more about himself from a biological viewpoint in order that he understand the nature of the problems with which he is confronted within himself and in relation to others. This could guide and provide a basis for judgment and for choosing and establishing patterns of behavior in the unfolding of each individual life so that each might then find more and increasing satisfaction in life and with life.

In these respects man is quite unfinished, both in his understanding and in his development. He has, obviously, a long way to go, but this is one way to proceed. The value will differ for each individual. Some may see hope, some may feel despair, while others may be totally uncomprehending. This, too, reveals the nature of the problem of man and reflects the unevenness with which man has evolved. The concept of unfinished and unfolding man is intended to convey the idea that the future *can* be different and that man's consciousness needs to be enhanced if he is to help guide himself individually and collectively toward the better life that he constantly seeks.

There are unexplored spaces within man and between men which are disordered or as yet unordered. There is an unfulfilled sense of order, or of harmony with nature, which also remains unsatisfied, incomplete, and unfinished.

When I speak of the wholeness of man, I have in mind the idea that the whole is greater than the sum of the parts and that the senses of man are related to the architecture of living matter, which produces effects created by the interrelationship of parts, and that these exist not by virtue of the parts alone but by virtue of their relationship.

To be fully functioning, it is necessary that the senses of the individual, from the earliest stages in life, be kept free of distortions and astigmatisms. The individual may then discover his own self, to which he is most intimately related, and find his way to other selves, with whom he can then become intimately related, by virtue of the natural compatibility and harmony that are revealed when such discoveries are made.

However, man is filled with preconceptions from which he must be freed so that he may see life as an unfolding process. He will then be able to decode the signals with which he is constantly bombarded and know what is right for him and for others who may be seeking him for the relationships that they, too, need. If one is not true to one's self in this respect, if one does not reveal the truth of his own being, then it is not possible to be true to others. Perhaps this is what is meant by young people when they speak of being "real" as compared to "unreal." This must be sensed before it can be known, and therefore man's senses, sensitivities, and sensibilities in these realms must be awakened to full consciousness. Then he would understand what is revealed to him upon seeing the truth in and about himself so that he may be faithful to himself and to others as well.

It is difficult to think in these terms in the presence of such overt pathology as war, or of conflicts made manifest in violence. While violence may be wholly pathological in origin, it may also be a manifestation of resistance and of an attempt to correct unhealthy extremes of excess or of greed inimical to the health of the species and therefore of individuals as well. When violence of this kind erupts, or before it erupts, a consideration of multiple-win rather than win-lose resolutions is more likely to reduce opposition to nonviolent compromises—resolutions in which neither side loses all and both sides gain some advantages in the transaction needed to establish a healthier equilibrium in the evolutionary scheme.

If man wants to continue his evolution in a healthful way, it will be necessary to reduce the amount of individual as well as collective pathology, and, at the same time, begin to develop ways and means of maintaining and enhancing health. The maintenance and enhancement of health will have the effect of suppressing or reducing pathology and therefore violence. The amount of resources, energy, and time available for enhancing health is inversely proportional to the amount required for dealing with pathology. As it becomes possible to reduce the resources, time, and energy required for dealing with pathology, those available for dealing with health will be augmented. In the process of improving the human condition it is necessary to give increasing attention to health while continuing to reduce pathology. If health is used as a means for suppressing or treating disease, then with increased resources and attention devoted to health, problems of disease would diminish and a chain reaction be set in motion that would accelerate the augmentation of health and the reduction of disease. This requires a totally different orientation than presently prevails.

Man needs to become increasingly conscious of what he is a part of so that he may then function fully, though individually as an element of the whole, in a more healthful way. This requires getting to know oneself so that one can be true to oneself and to others, if more harmony is to develop, and more order is to appear out of the conflict that is so manifest in the human condition at the present time. The thought that man is unfolding, in a developmental and evolutionary sense, provides a measure of hope for those who would work toward a more satisfying life for man on earth.

About the Author

Jonas Salk is presently Director and Fellow of the Salk Institute for Biological Studies in La Jolla, California. He is also an Adjunct Professor in the Health Sciences at the University of California, San Diego. After his studies at the College of the City of New York, he trained in medicine at the New York University and at Mount Sinai Hospital in New York City. He was drawn to research in various fields of biology, medicine, and public health, and at the University of Michigan and the University of Pittsburgh was involved in the development of means for the prevention of influenza and poliomyelitis. For a decade and a half Dr. Salk has been concerned with the relationship and application of fundamental biological discoveries, not only to solving problems of disease but to questions of health in man. The Institute for Biological Studies, which he founded, is dedicated not only to experimental biology but to relating biological knowledge to philosophical, psychological, and social questions. Dr. Salk views biology not only as a science but as a basic cultural discipline with unifying potential for the relationships that exist between man and the physical universe, as well as between man and the sciences, arts, and humanities.

About the Editor of This Series

Ruth Nanda Anshen, philosopher and editor, plans and edits *World Perspectives, Religious Perspectives, Credo Perspectives, Perspectives in Humanism* and *The Science of Culture Series.* She also writes and lectures on the relationship of knowledge to the nature and meaning of man and existence. Dr. Anshen's book, *The Reality of the Devil: Evil in Man,* is published by Harper & Row.

72 73 74 75 10 9 8 7 6 5 4 3 2 1